# The Theory of Optimal Stopping

*by*

Y. S. Chow
*Columbia University*

Herbert Robbins
*Columbia University and Rutgers University*

*and*

David Siegmund
*Stanford University*

DOVER PUBLICATIONS, INC.
New York

Published in Canada by General Publishing Company, Ltd., 30 Lesmill Road, Don Mills, Toronto, Ontario.
Published in the United Kingdom by Constable and Company, Ltd., 3 The Lanchesters, 162–164 Fulham Palace Road, London W6 9ER.

This Dover edition, first published in 1991, is an unabridged, corrected republication of the work first published by the Houghton Mifflin Company, Boston, 1971, under the title *Great Expectations: The Theory of Optimal Stopping*.

Manufactured in the United States of America
Dover Publications, Inc., 31 East 2nd Street, Mineola, N.Y. 11501

*Library of Congress Cataloging-in-Publication Data*

Chow, Yuan Shih, 1924–
The theory of optimal stopping / by Y.S. Chow, Herbert Robbins, and David Siegmund.
p. cm.
Rev. ed. of: Great expectations. 1971.
Includes bibliographical references and index.
ISBN 0-486-66650-6
1. Optimal stopping (Mathematical statistics) I. Robbins, Herbert. II. Siegmund, David, 1941– . III. Chow, Yuan Shih, 1924– . Great expectations. IV. Title.
QA279.7.C48 1991
519.2'87—dc20

90-19109
CIP

# Preface

This monograph arose from courses given during the past decade by the several authors on the subject of optimal stopping theory. It attempts to present systematically the mathematical foundations of the theory in discrete time together with a number of examples which illustrate the applicability of the general theory to particular problems. It is by no means a comprehensive survey of the field. For example, the reader will find no mention of continuous time problems (see Shiryaev [3]), nor will he find a proof of the optimal character of the sequential probability ratio test (see Ferguson [1]).

Particular optimal stopping problems, such as the "secretary problem" of Chapter 3 have a long history in probability theory. However, the first results of some generality were obtained in the late 1940's by Wald and Wolfowitz [1] and by Arrow, Blackwell, and Girshick [1], who were studying sequential statistical decision problems. A different approach to these problems was suggested by Snell [1] in 1952—see Section 4.4. Optimal stopping theory as a part of probability theory, with particular but not exclusive application to statistics, has been in a state of rapid development since about 1960, and it is mainly the authors' own contribution to this development which this monograph attempts to summarize.

The general theory of optimal stopping should be accessible to anyone with a knowledge of measure-theoretic probability, including some familiarity with the abstract notion of conditional expectation. A statement of basic prerequisites is presented in Chapter 1. Chapter 2 contains a brief exposition of martingale theory, the methods of which are similar to those of optimal stopping theory. The general theory of Chapters 3–5 does not depend on the results of Chapter 2 although the application of the theory to specific problems occasionally does. Hence the reader may wish to skim through Chapter 1 and go directly to Chapter 3, referring back to Chapter 2 only when necessary.

We would like to thank the numerous colleagues and students who through their interest and participation have assisted in bringing this project to its present state. Herman Chernoff has made many helpful suggestions in preparing the final text. We would also like to thank Miss Margaret Lof for an excellent job of typing.

Incomplete and imperfect as it is, this book is our tribute to a subject which has fascinated us for many years.

# Table of Contents

# Introduction

The theory of probability began with efforts to calculate the odds in games of chance. In this context, optimal stopping problems concern the effect on a gambler's fortune of various possible systems for deciding when to stop playing a sequence of games. A more recent field of application is statistical inference, where the experimenter must constantly ask whether the increase in information contained in further data will outweigh the cost of collecting it.

Optimal stopping theory provides a general mathematical framework in which such problems can be precisely formulated and in some cases solved completely. In this introduction we shall present four simple examples which illustrate some of the problems that arise in the general theory.

### Example 1 Favorable Modification of "Double or Nothing"

A fair coin is tossed repeatedly. After each toss we have the option of stopping or going on to the next toss, our decision at each stage being allowed to depend on the outcomes thus far. We must stop after some finite (but not preassigned) number of tosses, and it is agreed that if we stop after the $n$th toss we are to receive a reward $x_n$. In "Double or Nothing," for the player starting with 1 unit, the reward would be $\prod_{i=1}^{n} (y_i + 1)$ where $y_i = 1$ denotes heads on the $i$th toss and $y_i = -1$ tails. We, however, consider the case in which we are encouraged to continue playing by multiplying the above reward by the increasing sequence $2n/(n + 1)$, yielding the reward sequence

$$(1) \qquad x_n = \frac{2n}{n+1} \prod_{i=1}^{n} (y_i + 1) \qquad (n = 1, 2, \ldots),$$

which is 0 if there are any tails in the first $n$ tosses and $2^{n+1}n/(n+1)$ otherwise. This is a special case of the general situation where the reward at stage $n$, $x_n$, is any given function of the past, $x_n = f_n(y_1, y_2, \ldots, y_n)$.

A *stopping rule* is a random variable $t$ with values in the set $\{1, 2, 3, \ldots\}$ and such that the event $\{t = n\}$ depends solely on the values of $y_1, \ldots, y_n$ and not on future values $y_{n+1}, \ldots$. Using any stopping rule $t$, our reward $x_t$ will be a random variable whose expectation $Ex_t$ measures the performance on the average of the stopping rule $t$. The supremum $V = \sup \{Ex_t\}$ over the class $C$ of all possible stopping rules $t$ for which $Ex_t$ exists will be called the *value* of the sequence $\{x_n\}$, and if a particular stopping rule $t$ exists such that $Ex_t = V$, $t$ is said to be *optimal*.

Our general aim is to obtain methods of characterizing the value $V$ and an optimal rule if it exists. A little reflection will show that for the special case (1) we need consider only the class of stopping rules $\{t_k\}$, $k = 1, 2, \ldots$, where $t_k = k$; i.e., $t_k$ stops after the $k$th toss no matter what sequence of heads and tails has appeared. Clearly,

$$Ex_{t_k} = \frac{1}{2^k} \cdot \frac{k2^{k+1}}{k+1} + \left(1 - \frac{1}{2^k}\right) \cdot 0 = \frac{2k}{k+1},$$

and therefore $V = 2$ but no optimal stopping rule exists.

We remark in passing that at any stage $n$ in which only heads have appeared, so that $x_n = n2^{n+1}/(n+1)$, the conditional expected reward if we make just one more toss and then stop is

$$\begin{aligned} E(x_{n+1} \mid x_n = n2^{n+1}/(n+1)) &= \frac{1}{2} \frac{(n+1)2^{n+2}}{(n+2)} \\ &= 2^{n+1}(n+1)/(n+2) > x_n, \end{aligned}$$

so that it is always "foolish" to stop with all heads. That is, from the point of view of *expectation*, it is always better to toss once more. But if we do not act "foolishly" at some point we shall go on until the first tail occurs, an event which ultimately occurs with probability one, and our final reward will always be 0. Thus in this case acting "wisely" at each stage is the worst long-range policy. Incidentally it would be natural to regard this worst policy as a limiting form of the stopping rules $t_k$ whose expected rewards converge to the value $V = 2$.

This example illustrates a problem and two pathologies:

(a) An optimal rule may fail to exist. Under what conditions does an optimal rule exist and when can it be described simply?

(b) A policy which consists of acting "wisely" at each stage can be poor.

(c) A limiting version of a sequence of good policies can be poor.

Under what conditions can we be sure that the last two situations will not occur?

If we replace $x_n$ by $x'_n = 1 - x_n$, we obtain a sequence of *un*favorable games. Then the policy of waiting for a tail illustrates that

(d) A player may be able to assure himself a profit even though he plays a sequence of unfavorable games.

### Example 2 Reward Equal to Average

Leaving the $y$'s as in Example 1, let the reward be

$$x_n = \frac{y_1 + y_2 + \cdots + y_n}{n} \qquad (n = 1, 2, \ldots). \tag{2}$$

This problem is much harder than the preceding one because of the enormous family of candidates for stopping rules which must be considered. A simple instance of such a rule is

$$t = \begin{cases} 1 & \text{if} \quad y_1 = 1, \text{ otherwise} \\ n & \text{if} \quad n \text{ is the first integer such that } y_1 + \cdots + y_n = 0. \end{cases} \tag{3}$$

It is known (cf. problem 2.2) that for any integer $k$ the event $\{y_1 + \cdots + y_n = k \text{ for some } n\}$ occurs with probability 1, so $t$ is a legitimate stopping rule in the sense that $P(t < \infty) = 1$. Computing $Ex_t$ for (2) and (3) gives

$$Ex_t = \frac{1}{2} \cdot \frac{1}{1} + \frac{1}{2} \cdot 0 = \frac{1}{2}.$$

It follows that $V \geq \frac{1}{2}$, and of course $V \leq 1$, since $x_n \leq 1$ always. By trial and error we can invent other stopping rules $t$ for which $Ex_t > \frac{1}{2}$, but we shall find none for which $Ex_t > 0.9$, for example. However, it is not easy to *prove* that $V \leq 0.9$, still less to find the exact value of $V$ and to determine whether an optimal $t$ exists.

In cases like this it is tempting to try to "put the problem on a computer." Here is what a computer can do. Suppose we restrict ourselves to the class $C^N$ of all possible stopping rules $t$ which take on only values in the set $\{1, 2, \ldots, N\}$, where $N$ is some fixed positive integer; in other words, we restrict ourselves to stopping rules which always stop after at most $N$ tosses of the coin. Denoting by $V^N$ the supremum of $Ex_t$ over the class $C^N$, we see that because there are only a finite number of vectors $(y_1, \ldots, y_N)$ with $y_i = \pm 1$, $C^N$ is a finite class and therefore an optimal rule in $C^N$ must exist.

Even so, for $N = 1000$, say, the class $C^N$ is much too large for a direct analysis. At this point the general theory of optimal stopping comes to our aid. Wherever the number of stages in the problem is bounded (even if the random variables involved have continuous rather than discrete distributions) an optimal rule always exists, along with a more or less constructive algorithm for finding it (see Theorem 3.2). The general idea behind this algorithm may be summarized by the term "backward induction." In the present problem a computer program can be written to find $V^N$ quite quickly for all $N$ up to a few thousand. By definition the sequence $V^N$ is non-decreasing, and therefore $V' = \lim_{N\to\infty} V^N$ exists. But the values of $V^N$ produced by a computer for successive values of $N$ display no obvious pattern, and it is impossible to guess the exact value of $V'$ from computer evidence, *still less to decide whether* $V = V'$ *or* $V > V'$ *and whether an optimal* $t$ *exists in the class* $C$.

In the present problem it can in fact be proved (Theorems 4.3 and 4.11) that $V = V'$ and that an optimal $t$ does exist, but the numerical values of $V$ and the exact description of $t$ are not known.

In general we may ask

(e) Is it always true that $V = V'$?

(f) Are there effective ways of enclosing $V$ between computable upper and lower bounds which are close to each other?

### Example 3 Modified Sum

With the same $y$'s as in Example 1, let

$$x_n = \min(1, y_1 + \cdots + y_n) - \frac{n}{n+1} \qquad (n \geq 1).$$

Consider the particular stopping rule

$$(4) \qquad t = \text{first } n \geq 1 \quad \text{such that} \quad y_1 + \cdots + y_n = 1.$$

That $P(t < \infty) = 1$ follows as in Example 2, and since $\frac{n}{n+1} < 1$,

$$Ex_t = 1 - E\left(\frac{t}{t+1}\right) > 0$$

(the exact value of $Ex_t$ can be computed but need not concern us here). Hence $V > 0$, and a little thought will show that $t$ is in fact optimal for this example.

On the other hand, since the $y$'s are independent and identically distributed with $Ey_i = \frac{1}{2}\cdot 1 + \frac{1}{2}\cdot(-1) = 0$, Wald's lemma (cf.

Theorem 2.3 or Lemma 3.1) shows that if, unlike (4), $t$ is any stopping rule for which $Et < \infty$, and hence in particular if $t \in C^N$ for some $N = 1, 2, \ldots$, then $E(y_1 + \cdots + y_t) = 0$, so that

$$Ex_t \leq E(y_1 + \cdots + y_t) - E\left(\frac{t}{t+1}\right) \leq -\tfrac{1}{2}.$$

Hence $V' = \lim_{N\to\infty} V^N \leq -\frac{1}{2}$, while as we have seen $V > 0$. Thus the answer to (e) is No, but one may ask

(e′) Under what conditions is $V = V'$? In other words, when can the value be approximated by using procedures with a preassigned maximum number of stages?

A slight variation of this example is given by

**Example 4 Another Modified Sum**

Let $y_1, y_2, \ldots$ be independent (but not identically distributed) random variables such that

$$(5) \qquad P(y_i = 1 - a_i) = P(y_i = -1 - a_i) = \tfrac{1}{2}, \qquad \text{with } a_i = \frac{1}{i(i+1)},$$

and let

$$(6) \qquad x_n = y_1 + \cdots + y_n \qquad (n \geq 1)$$

be the reward if we stop at the $n$th stage. Thus $x_n$ represents the net gain of a gambler who plays a succession of unfavorable games,

$$Ey_i = \tfrac{1}{2}(1 - a_i) + \tfrac{1}{2}(-1 - a_i) = -a_i < 0,$$

and stops after the $n$th. It might seem that whatever stopping rule $t$ he might use, his expected net gain after stopping, $Ex_t$, would be $< 0$. However, let

$$(7) \qquad t = \text{first } n \geq 1 \quad \text{such that} \quad \sum_1^n (y_i + a_i) = k,$$

where $k$ is any preassigned positive integer. The argument in Example 2 again shows that $P(t < \infty) = 1$, while by (6) and (7),

$$Ex_t = E\left(\sum_1^t y_i\right) = k - E\left(\sum_1^t a_i\right) = k - E\left(\sum_1^t \left(\frac{1}{i} - \frac{1}{i+1}\right)\right)$$

$$= k - E\left(\frac{t}{t+1}\right) > k - 1.$$

Since $k$ can be as large as we please, we see that $V = +\infty$ for this example. (The reader may decide whether there is an optimal $t$ in this case, i.e., one for which $Ex_t = +\infty$.) We remark that although in this example the one-step conditional expected reward $E(x_{n+1} \mid x_n) = x_n + E(y_{n+1})$ is always *less* than the present reward $x_n$, the use of a proper stopping rule makes the game very profitable.

# Chapter 1

# Preliminaries

This chapter summarizes some standard results in measure theory and integration theory which will be used in subsequent chapters.

## 1. Algebras of Events

A class $\mathscr{F}$ of subsets of a set $\Omega$ is called an *algebra* if

(1.1) $$\Omega \in \mathscr{F}$$

(1.2) $$\bar{A} \equiv \Omega - A \in \mathscr{F} \quad \text{and} \quad A \cup B \in \mathscr{F} \quad \text{whenever} \quad A \in \mathscr{F}, B \in \mathscr{F}.$$

$\mathscr{F}$ is called a $\sigma$-algebra if in addition to (1.1) and (1.2),

(1.3) $$\bigcup_1^\infty A_n \in \mathscr{F} \qquad \text{whenever } A_n \in \mathscr{F} \text{ for all } n = 1, 2, \ldots.$$

The sets $A$ in a $\sigma$-algebra $\mathscr{F}$ are called *events*. The class of all subsets of $\Omega$ is clearly a $\sigma$-algebra, as is the class consisting of the empty set $\phi$ and $\Omega$. The former class is the largest $\sigma$-algebra of subsets of $\Omega$; the latter class is the smallest.

Given any class $\mathscr{A}$ of subsets of $\Omega$, the intersection of all $\sigma$-algebras containing $\mathscr{A}$ is itself a $\sigma$-algebra containing $\mathscr{A}$, and this intersection is contained in every $\sigma$-algebra which contains $\mathscr{A}$. It is called the $\sigma$-algebra *generated by* $\mathscr{A}$ and is denoted by $\mathscr{B}(\mathscr{A})$. If $\mathscr{A}$ is the class of intervals of the extended real line $[-\infty, \infty]$, then $\mathscr{B}(\mathscr{A})$ is called the $\sigma$-algebra of (*linear*) *Borel sets*, and is denoted by $\mathscr{B}$.

## 2. Random Variables

Given sets $\Omega, X$ with respective $\sigma$-algebras $\mathscr{F}, \mathscr{X}$, a function $x: \Omega \to X$ is called a *random variable* (r.v.) if $x^{-1}(A) \in \mathscr{F}$ for every $A \in \mathscr{X}$. When several $\sigma$-algebras of subsets of $\Omega$ are under consideration, to avoid ambiguity we sometimes say that $x$ is an $\mathscr{F}$-measurable r.v. If $X = [-\infty, \infty]$ and $\mathscr{X} = \mathscr{B}$, then $x$ is called a real random variable (r.r.v.) on $\Omega$. If $B \in \mathscr{F}$, $I_B$ (the indicator of $B$) denotes the r.r.v. which is 1 on $B$ and 0 off $B$. If $\{x_t, t \in T\}$ is a family of random variables with $x_t$: $\Omega \to X_t$ then $\mathscr{B}(x_t, t \in T)$ denotes the smallest $\sigma$-algebra with respect to which the r.v.'s $x_t, t \in T$ are all measurable and is called the $\sigma$-algebra generated by $\{x_t, t \in T\}$.

## 3. Probabilities and Expectations

A non-negative extended real-valued set function $\mu$ on an algebra $\mathscr{F}$ of subsets of a set $\Omega$ is called a *measure* if $\mu(\phi) = 0$ and

$$\mu\left(\bigcup_1^\infty A_n\right) = \sum_1^\infty \mu(A_n) \tag{1.4}$$

whenever $A_n \in \mathscr{F}$ $(n \geq 1)$, $A_n \cap A_m = \phi$ $(n \neq m)$ and $\bigcup_{n=1}^\infty A_n \in \mathscr{F}$. The measure $\mu$ is said to be *$\sigma$-finite* if $\Omega$ is a denumerable union of sets in $\mathscr{F}$ of finite measure; it is called a *probability* measure if

$$\mu(\Omega) = 1. \tag{1.5}$$

As a rule probability measures will be denoted by the letter $P$. If $\mathscr{F}$ is a $\sigma$-algebra and $P$ is a probability measure, the triple $(\Omega, \mathscr{F}, P)$ is called a *probability space*.

**Theorem 1.1 Extension Theorem** Given an algebra $\mathscr{A}$ of subsets of a set $\Omega$, and a probability measure $P$ on $\mathscr{A}$, there exists a unique probability measure on $\mathscr{B}(\mathscr{A})$ which coincides with $P$ on $\mathscr{A}$.

The following result, which proves the uniqueness part of the extension theorem, is of independent interest.

**Lemma 1.1** Given a probability space $(\Omega, \mathscr{F}, P)$ and an algebra $\mathscr{A}$ such that $\mathscr{F} = \mathscr{B}(\mathscr{A})$, for each $A \in \mathscr{F}$ and $\varepsilon > 0$ there exists a $B \in \mathscr{A}$ such that

$$P(A - B) + P(B - A) < \varepsilon. \tag{1.6}$$

***Proof*** Let $\mathscr{C}$ denote the class of all $A \in \mathscr{F}$ with this property. Clearly $\mathscr{C} \supset \mathscr{A}$, and it is easy to check that $\mathscr{C}$ is a $\sigma$-algebra. Hence $\mathscr{C} \supset \mathscr{F}$.

Let $(\Omega, \mathscr{F}, P)$ be a probability space and $x: \Omega \to X$ a random variable. The *image* of $P$ under $x$ is the probability measure $Q$ on $\mathscr{X}$

defined by $Q(A) = P(x^{-1}(A))$ for all $A \in \mathscr{X}$. $Q$ is also called the *distribution* of $x$.

Let $(\Omega, \mathscr{F}, P)$ be a probability space and $x$ a non-negative r.r.v. The *expectation* of $x$, denoted by $Ex$, is defined as

$$Ex = \lim_{n\to\infty} \left( \sum_{k=0}^{n2^n} k2^{-n}P\{k2^{-n} < x \leq (k+1)2^{-n}\} + nP\{x > n\} \right).$$

(By $\{\cdots\}$ we mean "the set of all points $\omega$ in $\Omega$ such that . . .".) If $x$ is any r.r.v., not necessarily non-negative, for which either $Ex^+ < \infty$ or $Ex^- < \infty$, where by definition $x^+ = \max(x, 0)$, $x^- = -\min(x, 0)$, then we say that $Ex$ exists and define $Ex = Ex^+ - Ex^-$. For $E(I_A x)$ we frequently write $\int_A x$ or $\int_A x\,dP$. Elementary properties are the following:

(1.7) $$Ex \geq 0 \quad \text{if} \quad x \geq 0$$

(1.8) $$E1 = 1$$

(1.9) If $Ex$ exists, then $E(cx)$ exists and $E(cx) = cEx$ for any real number $c$.

(1.10) If $Ex_1$ and $Ex_2$ exist, then $E(x_1 + x_2)$ exists and equals $Ex_1 + Ex_2$ provided the latter expression is not of the form $+\infty - \infty$.

**Theorem 1.2 Monotone Convergence Theorem** If $x_n \uparrow x$ and $Ex_1^- < \infty$, then $Ex_n \uparrow Ex$. Similarly, if $x_n \downarrow x$ and $Ex_1^+ < \infty$, then $Ex_n \downarrow Ex$.

***Remark*** The conclusion of the theorem is unchanged if we merely suppose that $P\{x_n \uparrow x\} = 1$, i.e., that $x_n \uparrow x$ *with probability one* or *almost surely* (a.s.). In what follows convergence of a sequence of r.r.v.'s on a probability space, unless otherwise specified, will be understood to mean almost sure convergence. Similarly, statements like "$x \leq y$" and "$x_n \to x$ on $A$" should be interpreted as "$P\{x \leq y\} = 1$" and "$P(A \cap \{x_n \nrightarrow x\}) = 0$."

## 4. Uniform Integrability

A sequence $(x_n)$ of r.r.v.'s is *uniformly integrable* if

(1.11) $$\lim_{a\to\infty} \sup_n \int_{\{|x_n| > a\}} |x_n| = 0$$

or equivalently if

(1.12) $$\sup_n E|x_n| < \infty \quad \text{and} \quad \lim_{P(A)\to 0} \sup_n \int_A |x_n| = 0.$$

**Lemma 1.2 Fatou's Lemma** If $(x_n^+)$ is uniformly integrable and $E(\limsup_n x_n)$ exists, then

$$E(\limsup_n x_n) \geq \limsup_n Ex_n.$$

***Proof*** If $x_n \leq a < \infty$ for all $n \geq 1$, we observe that

$$\sup_{k \geq n} x_k \downarrow \limsup_n x_n \quad \text{as} \quad n \to \infty,$$

and

$$E\left(\sup_{k \geq 1} x_k\right) \leq a < \infty.$$

Therefore we may apply the second part of Theorem 1.2 to get

$$E(\sup_{k \geq n} x_k) \downarrow E(\limsup_n x_n).$$

This together with the fact that $E(\sup_{k \geq n} x_k) \geq Ex_n$ completes the proof in the special case. In the general case, let $\varepsilon > 0$ and $x_n(a) = \min(x_n, a)$, where $a$ is so large that $Ex_n(a) \geq Ex_n - \varepsilon$ $(n \geq 1)$. Then

$$E(\limsup_n x_n) \geq E(\limsup_n x_n(a)) \geq \limsup_n Ex_n - \varepsilon.$$

Letting $\varepsilon \to 0$ completes the proof.

**Theorem 1.3** Assume that $0 \leq x_n \to x$ and $Ex_n < \infty$ $(n \geq 1)$. Then $Ex_n \to Ex < \infty$ if and only if $(x_n)$ is uniformly integrable.

***Proof*** To prove the "if" part, note that Fatou's lemma applied to $(x_n)$ and to $(-x_n)$ yields

$$Ex \leq \liminf_n Ex_n \leq \limsup_n Ex_n \leq Ex. \tag{1.13}$$

(The first part of (1.12) and (1.13) show that $Ex < \infty$.) To prove the "only if" part, first observe that $B = \{b : P\{x = b\} > 0\}$ is countable. (In fact for each $m = 1, 2, \ldots,$ $\left\{b : P\{x = b\} \geq \frac{1}{m}\right\}$ is finite.) For any $a \notin B$

$$x_n I_{\{x_n < a\}} \to x I_{\{x < a\}},$$

and obviously $(x_n I_{\{x_n < a\}})$ satisfies the first definition of uniform integrability.

Hence by the "if" part $\int_{\{x_n<a\}} x_n \to \int_{\{x<a\}} x$. Thus $\int_{\{x_n \geq a\}} x_n \to \int_{\{x_n \geq a\}} x$ for each $a \notin B$. Let $\varepsilon > 0$ and choose $a_0 \notin B$ so large that $\int_{\{x \geq a_0\}} x < \varepsilon/2$. Let $N_0$ be so large that for all $n \geq N_0$

$$\int_{\{x_n \geq a_0\}} x_n \leq \int_{\{x \geq a_0\}} x + \varepsilon/2,$$

and finally let $a_1 \geq a_0$ be so large that $\int_{\{x_n \geq a_1\}} x_n < \varepsilon$ for all $n = 1, 2, \ldots, N_0$. Then for all $a \geq a_1$ we have

$$\int_{\{x_n > a\}} x_n < \varepsilon \qquad (n = 1, 2, \ldots).$$

**Corollary 1** If $x_n \to x$ and $(x_n)$ is uniformly integrable, then $E|x_n - x| \to 0$.

***Proof*** By Fatou's lemma and the first part of (1.12) $E|x| \leq \liminf E|x_n| < \infty$. It is easily seen from (1.12) that $(|x_n - x|)$ is uniformly integrable, and since $|x_n - x| \to 0$, we have by Theorem 1.3 $E|x_n - x| \to 0$.

**Corollary 2 Lebesgue Dominated Convergence Theorem** If $x_n \to x$ and if there exists a random variable $y$ such that $|x_n| \leq y$ $(n \geq 1)$ and $Ey < \infty$, then $E|x_n - x| \to 0$.

***Remark*** We say that $x_n$ converges to $x$ *in probability* and write $x_n \xrightarrow{P} x$ if for every $\varepsilon > 0$, $\lim_{n\to\infty} P\{|x_n - x| > \varepsilon\} = 0$. Using the fact that $\{x_n \to x\} = \bigcap_{r=1}^{\infty} \bigcup_{n=1}^{\infty} \bigcap_{k=n}^{\infty} \{|x_k - x| < r^{-1}\}$, it is easy to see that if $x_n \to x$, then $x_n \xrightarrow{P} x$. The converse is not true, but Theorem 1.3 and its corollaries remain true if the assumption $x_n \to x$ is replaced by $x_n \xrightarrow{P} x$. Since the notion of convergence in probability plays a minor role in what follows, we omit the details.

A fundamental result in arguments involving conditional expectations (to be defined below) is

**Lemma 1.3** If $Ex_1$ and $Ex_2$ exist and for each $A \in \mathscr{F}$

$$\int_A x_1 \leq (=) \int_A x_2$$

then $x_1 \leq (=) x_2$.

***Proof*** For each $m > 0$

$$\int_{A\{|x_1| \leq m\}} (x_1 - x_2) \leq 0.$$

Setting $A = \{x_1 - x_2 > \varepsilon\}$, we obtain

$$P\{|x_1| \leq m, x_1 - x_2 > \varepsilon\} = 0.$$

Letting $\varepsilon \to 0$, $m \to \infty$, we have

$$P\{|x_1| < \infty, x_1 > x_2\} = 0,$$

and by similar arguments

$$P\{|x_2| < \infty, x_1 > x_2\} = 0.$$

Since $Ex_1 \leq Ex_2$ implies that $P\{x_1 = +\infty, x_2 = -\infty\} = 0$, we conclude that $P\{x_1 \leq x_2\} = 1$. The case of equality follows at once from the preceding case when the roles of $x_1$ and $x_2$ are interchanged.

## 5. Conditional Expectations

Given a sub-$\sigma$-algebra $\mathscr{G}$ of $\mathscr{F}$ and a non-negative r.r.v. $x$, "the" *conditional expectation of $x$ given $\mathscr{G}$* is any $\mathscr{G}$-measurable r.r.v. $g$ such that

$$\int_A x = \int_A g \quad \text{for every} \quad A \in \mathscr{G}.$$

By Lemma 1.3 such a $g$ is uniquely determined up to a set of probability 0, and the existence of such a $g$ is guaranteed by the Radon-Nikodym theorem. If $x$ is any r.r.v. for which $Ex$ exists, then the conditional expectation of $x$ given $\mathscr{G}$ (denoted by $E(x \mid \mathscr{G})$) is defined as

$$E(x \mid \mathscr{G}) = E(x^+ \mid \mathscr{G}) - E(x^- \mid \mathscr{G}).$$

If $x_0, x_1, \ldots$ are r.r.v.'s such that $Ex_n$ exists for all $n \geq 0$, then (up to an event of probability 0)

(1.14) $$E(x_0 \mid \mathscr{G}) \geq 0 \quad \text{if} \quad x_0 \geq 0.$$

(1.15) $$E(1 \mid \mathscr{G}) = 1.$$

(1.16) $E(x_0 + x_1 \mid \mathscr{G}) = E(x_0 \mid \mathscr{G}) + E(x_1 \mid \mathscr{G})$ provided $Ex_0 + Ex_1$ is not of the form $+\infty - \infty$.

(1.17) $E(x_0 x_1 \mid \mathscr{G}) = x_0 E(x_1 \mid \mathscr{G})$ whenever $x_0$ is $\mathscr{G}$-measurable and $E(x_0 x_1)$ exists.

(1.18) If $\mathscr{G}^* \subset \mathscr{G}$, then $E(x_0 \mid \mathscr{G}^*) = E(E(x_0 \mid \mathscr{G}) \mid \mathscr{G}^*)$.

(1.19) Monotone Convergence Theorem for Conditional Expectations: If $x_n \uparrow x$ and $Ex$ exists, then $E(x \mid \mathscr{G}) = \lim_{n\to\infty} E(x_n \mid \mathscr{G})$ on $\{E(x_0 \mid \mathscr{G}) > -\infty\}$.

(1.20) Fatou's Lemma for Conditional Expectations: If $x_n \leq x_0$ $(n \geq 0)$ and $E(\limsup x_n)$ exists, then $E(\limsup x_n \mid \mathscr{G}) \geq \limsup E(x_n \mid \mathscr{G})$ on $\{E(x_0 \mid \mathscr{G}) < \infty\}$.

(1.21) Dominated Convergence Theorem for Conditional Expectations: If $|x_n| \leq x_0$ $(n \geq 1)$, $x_n \to x$, and $Ex$ exists, then $E(|x_n - x| \mid \mathscr{G}) \to 0$ on $\{E(x_0 \mid \mathscr{G}) < \infty\}$.

We shall indicate the proof of (1.19). (1.20) follows directly from (1.19) in the same way that Fatou's lemma follows from the monotone convergence theorem, and (1.21) is an immediate consequence of (1.20).

For any $m = 1, 2, \ldots,$ put $B_m = \{E(x_0 \mid \mathscr{G}) > -m\}$, $B = \bigcup_1^\infty B_m = \{E(x_0 \mid \mathscr{G}) > -\infty\}$. $x_n \uparrow x$ implies that $E(x_n \mid \mathscr{G}) \uparrow$. Let $y = \lim_{n\to\infty} E(x_n \mid \mathscr{G})$. Since

$$\int_{B_m} x_0 = \int_{B_m} E(x_0 \mid \mathscr{G}) \geq -m > -\infty,$$

we have by the monotone convergence theorem for each $A \in \mathscr{G}$ and $m = 1, 2, \ldots$

$$\int_{AB_m} y = \lim_n \int_{AB_m} E(x_n \mid \mathscr{G}) = \lim_n \int_{AB_m} x_n = \int_{AB_m} x = \int_{AB_m} E(x \mid \mathscr{G}).$$

It follows from Lemma 1.3 that

$$y = E(x \mid \mathscr{G}) \quad \text{on} \quad B_m,$$

and since $m$ is arbitrary

$$y = E(x \mid \mathscr{G}) \quad \text{on} \quad B.$$

**Theorem 1.4** **(P. Lévy)** Let $\{\mathscr{F}_n, n \geq 1\}$ be a non-decreasing sequence of sub-$\sigma$-algebras of $\mathscr{F}$, and put $\mathscr{F}_\infty = \mathscr{B}\left(\bigcup_1^\infty \mathscr{F}_n\right)$. For any r.r.v. $x$ with $E|x| < \infty$, $\lim_{n\to\infty} E(x \mid \mathscr{F}_n) = E(x \mid \mathscr{F}_\infty)$.

***Proof*** Putting $x' = E(x \mid \mathscr{F}_\infty)$, we see by (1.18) that without loss of generality we may assume that $x$ is $\mathscr{F}_\infty$-measurable, $x = E(x \mid \mathscr{F}_\infty)$. Let $\varepsilon > 0$. Since $\bigcup_1^\infty \mathscr{F}_n$ is an algebra, an easy consequence of Lemma 1.1 and the fact that (1.6) is equivalent to $E|I_A - I_B| < \varepsilon$ is that there exists an $n_0 = n_0(\varepsilon)$ and an $x_{n_0}$ which is $\mathscr{F}_{n_0}$-measurable such that

$$E|x - x_{n_0}| < \varepsilon^2/2.$$

Set $y = |x - x_{n_0}|$ and

$$\begin{aligned} t &= \text{first } n \geq n_0 \quad \text{for which } E(y \mid \mathscr{F}_n) > \varepsilon \\ &= \infty \quad \text{if no such } n \text{ exists.} \end{aligned}$$

Then since $\{t = n\} \in \mathscr{F}_n$ for all $n \geq n_0$, we may write

$$\begin{aligned} P\{E(y \mid \mathscr{F}_n) > \varepsilon \quad \text{for some } n \geq n_0\} &= \sum_{n=n_0}^{\infty} P\{t = n\} \\ &\leq \sum_{n_0}^{\infty} \frac{1}{\varepsilon} \int_{\{t=n\}} E(y \mid \mathscr{F}_n) = \sum_{n_0}^{\infty} \frac{1}{\varepsilon} \int_{\{t=n\}} y \leq \frac{1}{\varepsilon} Ey = \frac{\varepsilon}{2}. \end{aligned}$$

But for all $n \geq n_0$, since $x_{n_0}$ is $\mathscr{F}_n$-measurable,

$$\begin{aligned} |E(x \mid \mathscr{F}_n) - x| &\leq |E(x - x_{n_0} \mid \mathscr{F}_n)| + |x_{n_0} - x| \\ &\leq E(y \mid \mathscr{F}_n) + y, \end{aligned}$$

and hence $P\{|E(x \mid \mathscr{F}_n) - x| > 2\varepsilon \text{ for some } n \geq n_0\}$

$$\begin{aligned} &\leq P\{E(y \mid \mathscr{F}_n) > \varepsilon \text{ for some } n \geq n_0\} + P\{y > \varepsilon\} \\ &\leq \frac{\varepsilon}{2} + \frac{1}{\varepsilon} Ey = \frac{\varepsilon}{2} + \frac{\varepsilon}{2} = \varepsilon. \end{aligned}$$

## 6. Essential Supremum

Let $\{x_t, t \in T\}$ be a family of r.r.v.'s on a probability space $(\Omega, \mathscr{F}, P)$. We say that a r.r.v. $y$ is the essential supremum of $\{x_t, t \in T\}$ and write $y = \operatorname*{ess\,sup}_{t \in T} x_t$ if

(i) $\quad P\{y \geq x_t\} = 1 \quad \text{for every} \quad t \in T;$

(ii) if $y'$ is any r.r.v. such that $P\{y' \geq x_t\} = 1$ for every $t \in T$, then $P\{y' \geq y\} = 1$.

**Theorem 1.5** $y = \operatorname{ess\,sup}_{t\in T} x_t$ always exists, and for some countable subset $C$ of $T$,

$$y = \sup_{t\in C} x_t.$$

***Proof*** By passing to arc tan $x_t$ ($t \in T$) if necessary, we may assume that $|x_t(\omega)| \leq \pi/2$ for *all* $t \in T$ and $\omega \in \Omega$. Let $b$ denote the sup of

$$E(\sup_{t\in A} x_t),$$

as $A$ runs through the class of *finite* subsets of $T$. Let $C = \bigcup_1^\infty A_n$, where $A_n$ satisfies

$$E(\sup_{t\in A_n} x_t) \geq b - n^{-1}$$

and put $y = \sup_{t\in C} x_t$. $C$ is clearly countable, and it is easy to verify that $y$ has properties (i) and (ii).

## 7. Independent Random Variables and the Strong Law of Large Numbers

A finite family $\{\mathscr{F}_1, \ldots, \mathscr{F}_k\}$ of sub-$\sigma$-algebras of $\mathscr{F}$ are called *independent* if $P(A_1 \cap \cdots \cap A_k) = P(A_1)\cdots P(A_k)$ for every choice of $A_1, \ldots, A_k$ with $A_i \in \mathscr{F}_i$, $i = 1, 2, \ldots, k$. An arbitrary family $\{\mathscr{F}_t, t \in T\}$ is independent if every finite subfamily of it is independent. A family of r.v.'s $\{x_t, t \in T\}$ is called independent if the family of sub-$\sigma$-algebras $\{\mathscr{B}(x_t), t \in T\}$ is independent.

Two sub-$\sigma$-algebras $\mathscr{F}_1$ and $\mathscr{F}_2$ are independent if and only if $E(x \mid \mathscr{F}_1) = Ex$ for every bounded $\mathscr{F}_2$-measurable r.r.v. $x$. A family $\{\mathscr{F}_t, t \in T\}$ is independent if and only if for any two finite disjoint subsets $\{t_1, \ldots, t_n\}$ and $\{s_1, \ldots, s_m\}$ of $T$, $\mathscr{B}(\mathscr{F}_{t_1} \cup \cdots \cup \mathscr{F}_{t_n})$ and $\mathscr{B}(\mathscr{F}_{s_1} \cup \cdots \cup \mathscr{F}_{s_m})$ are independent.

**Theorem 1.6 Strong Law of Large Numbers** Let $x_1, x_2, \ldots$ be a sequence of independent, identically distributed (i.i.d.) real random variables. If $Ex_1$ exists, then

$$P\{\lim_{n\to\infty} n^{-1}(x_1 + \cdots + x_n) = Ex_1\} = 1. \tag{1.22}$$

If there exists a *finite* constant $c$ such that

$$P\{\lim_{n\to\infty} n^{-1}(x_1 + \cdots + x_n) = c\} = 1, \tag{1.23}$$

then $Ex_1$ exists and equals $c$.

***Proof*** If $Ex_1$ exists and is finite, (1.22) follows from Section 2.2(c). Suppose that $Ex_1 = +\infty$. (The case $Ex_1 = -\infty$ is treated similarly.) For each $a > 0$ there exists $g = g(a)$ such that

$$\int_{\{x_1 \le g\}} x_1 \ge a.$$

Define $x_n' = x_n I_{\{x_n \le g\}}$. Then $a \le Ex_n' \le g < +\infty$ and hence

$$\liminf_n n^{-1} \sum_1^n x_k \ge \lim_n n^{-1} \sum_1^n x_k' \ge a.$$

Since $a$ is arbitrary it follows that (1.22) holds in this case as well. Now suppose that (1.23) holds and hence $x_n/n \to 0$. It follows from the Borel-Cantelli lemma (see Section 2.4(d)) that

$$\sum_1^\infty P\{|x_1| > n\} = \sum_1^\infty P\{|x_n| > n\} < \infty$$

and hence $E|x_1| < \infty$. From the first part of the theorem it now follows that $Ex_1 = c$.

# Chapter 2

# Martingales. Wald's Lemma. Applications

This chapter establishes some basic theorems about martingales and will serve to acquaint the reader with reasoning which is characteristic of the optimal stopping theory developed in later chapters.

## 1. Definitions, Examples, Convergence Theorem

Let $(\Omega, \mathscr{F}, P)$ be a probability space, and let $I$ be any interval of the form $(a, b)$, $[a, b)$, $(a, b]$, or $[a, b]$ of the ordered set

$$\{-\infty, \ldots, -1, 0, 1, \ldots, +\infty\}.$$

By a *submartingale* we mean a family of pairs $\{x_n, \mathscr{F}_n, n \in I\}$ such that:

(i) $$\mathscr{F}_m \subset \mathscr{F}_n \subset \mathscr{F} \quad \text{for all} \quad m < n,$$

(ii) $x_n$ is an $\mathscr{F}_n$-measurable r.r.v. with $Ex_n^+ < \infty$ for all $n$,

(iii) $$x_m \leq E(x_n \mid \mathscr{F}_m) \quad \text{for all} \quad m < n$$

(or equivalently, $\int_A x_m \leq \int_A x_n$ for all $m < n$ and $A \in \mathscr{F}_m$). If $I$ contains neither $+\infty$ nor $-\infty$, then (iii) may be replaced by $x_n \leq E(x_{n+1} \mid \mathscr{F}_n)$ for all $n$ and $n+1 \in I$. If $\{-x_n, \mathscr{F}_n, n \in I\}$ is a submartingale, then $\{x_n, \mathscr{F}_n, n \in I\}$ is called a *supermartingale*. If $E|x_n| < \infty$ $(n \in I)$ and equality holds in (iii), i.e., if $\{x_n, \mathscr{F}_n, n \in I\}$ is simultaneously a submartingale and a supermartingale, then it is called a *martingale*.

## Examples

(a) Let $\{\mathscr{F}_n, n \in I\}$ satisfy (i) and let $z$ be any r.r.v. for which $E|z| < \infty$. Put $x_n = E(z \mid \mathscr{F}_n)$ $(n \in I)$. Then for any $m, n \in I$ $(m < n)$, we have by (1.18)

$$E(x_n \mid \mathscr{F}_m) = E[E(z \mid \mathscr{F}_n) \mid \mathscr{F}_m] = E(z \mid \mathscr{F}_m) = x_m,$$

and hence $\{x_n, \mathscr{F}_n, n \in I\}$ is a martingale.

(b) Let $y_1, y_2, \ldots$ be independent r.r.v.'s with expectation 0. Put $x_n = y_1 + \cdots + y_n$, $\mathscr{F}_n = \mathscr{B}(y_1, \ldots, y_n)$. Then $\{x_n, \mathscr{F}_n, 1 \le n < \infty\}$ is a martingale.

(c) Let $\{\mathscr{F}_n, 1 \le n < \infty\}$ satisfy (i). For any probability measure $Q$ on $\mathscr{F}_\infty \equiv \mathscr{B}\left(\bigcup_1^\infty \mathscr{F}_n\right)$ let $Q_n$ denote the restriction of $Q$ to $\mathscr{F}_n$, and suppose that $Q$ is such that each $Q_n$ is absolutely continuous with respect to the corresponding restriction $P_n$ of $P$, i.e., $Q_n(A) = 0$ for every $A \in \mathscr{F}_n$ for which $P_n(A) = 0$. Let $x_n = \dfrac{dQ_n}{dP_n}$ be the Radon-Nikodym derivative of $Q_n$ with respect to $P_n$. Then for any $A \in \mathscr{F}_n$ and any $m > n$

$$\int_A x_n = Q_n(A) = Q_m(A) = \int_A x_m,$$

and it follows that $\{x_n, \mathscr{F}_n, 1 \le n < \infty\}$ is a martingale. In particular if $y_1, y_2, \ldots$ are i.i.d. with probability density $f$ with respect to some $\sigma$-finite measure $\mu$ on the Borel sets of the line, and $g$ is any other probability density with respect to $\mu$ such that $\int_A g\, d\mu = 0$ for any linear Borel set $A$ for which $\int_A f\, d\mu = 0$, then with $\mathscr{F}_n = \mathscr{B}(y_1, \ldots, y_n)$,

$$\left\{\frac{g(y_1)\cdots g(y_n)}{f(y_1)\cdots f(y_n)}, \quad \mathscr{F}_n, 1 \le n < \infty\right\} \quad \text{is a martingale.}$$

If $\varphi(\lambda) \equiv E(e^{\lambda y_1}) < \infty$ for some non-zero (real) $\lambda$, we may take $\mu(\cdot) = P\{y_1 \in (\cdot)\}$ (and hence $f \equiv 1$) and $g(y) = e^{\lambda y}/\varphi(\lambda)$. Then $\{e^{\lambda(y_1 + \cdots + y_n)}/(\varphi(\lambda))^n, \mathscr{F}_n, 1 \le n < \infty\}$ is a martingale, as can easily be verified directly.

***Remark*** We have assumed for simplicity that $Q_n$ is absolutely continuous with respect to $P_n$ for every $n \ge 1$. More generally we may put $R = \frac{1}{2}(P + Q)$, $\tilde{x}_n = \dfrac{dQ_n/dR_n}{dP_n/dR_n}$. Then it is not hard to see that $\{\tilde{x}_n, \mathscr{F}_n, 1 \leq n < \infty\}$ is a supermartingale which equals $\{x_n, \mathscr{F}_n, 1 \leq n < \infty\}$ in the absolutely continuous case.

(d) Let $y_1, y_2, \ldots$ be i.i.d. with $E|y_1| < \infty$. For each $n = 1, 2, \ldots$ let $s_n = y_1 + \cdots + y_n$, $x_{-n} = \dfrac{s_n}{n}$, $\mathscr{F}_{-n} = \mathscr{B}(s_n, s_{n+1}, \ldots)$. Then, by symmetry, for any $n \geq 1$, $k = 1, 2, \ldots, n$

$$E(y_1 \mid \mathscr{F}_{-n}) = E(y_k \mid \mathscr{F}_{-n}),$$

and since $\sum_{k=1}^{n} E(y_k \mid \mathscr{F}_{-n}) = E(s_n \mid \mathscr{F}_{-n}) = s_n$, it follows that $E(y_1 \mid \mathscr{F}_{-n}) = \dfrac{s_n}{n} = x_{-n}$, and hence by (a) above that

$$\{x_n, \mathscr{F}_n, -\infty < n \leq -1\} \quad \text{is a martingale.}$$

(e) Let $y_1, y_2, \ldots$ be any sequence of random variables for which $E|y_k| < \infty$ for all $k \geq 1$. Put $\mathscr{F}_0 = \{\phi, \Omega\}$ and $\mathscr{F}_n = \mathscr{B}(y_1, \ldots, y_n)$. Then $\left\{\sum_1^n (y_k - E(y_k \mid \mathscr{F}_{k-1})), \mathscr{F}_n, 1 \leq n < \infty\right\}$ is a martingale. If $\{z_n, \mathscr{F}_n, 1 \leq n < \infty\}$ is a submartingale with $Ez_1^- < \infty$, then with

$$z_0 = 0, y_n = z_n - z_{n-1} \qquad (n \geq 1),$$

we may write $z_n = \sum_1^n (y_k - E(y_k \mid \mathscr{F}_{k-1})) + \sum_1^n E(y_k \mid \mathscr{F}_{k-1})$. Since by the submartingale property

$$E(y_k \mid \mathscr{F}_{k-1}) = E(z_k - z_{k-1} \mid \mathscr{F}_{k-1}) \geq 0 \qquad (k \geq 2)$$

we have the representation

$$z_n = x_n + \alpha_n, \tag{2.1}$$

where $\{x_n, \mathscr{F}_n, 1 \leq n < \infty\}$ is a martingale and $Ez_1 \leq \alpha_n \uparrow$.

**Lemma 2.1** If $\{x_n, \mathscr{F}_n, n \in I\}$ is a (sub) martingale, and if $\varphi$ is a real-valued (increasing) convex function of the real variable $a$ such that $E(\varphi(x_{n_0}))^+ < \infty$ for some $n_0 \in I$, then

$$\{\varphi(x_n), \mathscr{F}_n, n \in I, n \leq n_0\}$$

is a submartingale.

***Proof*** Let $m, n \in I$, $m \leq n \leq n_0$. Then by Jensen's inequality for conditional expectations (see Loeve [1], p. 348)

$$\varphi(x_m) \leq \varphi(E(x_{n_0} \mid \mathscr{F}_m)) \leq E(\varphi(x_{n_0}) \mid \mathscr{F}_m). \tag{2.2}$$

Thus $E(\varphi(x_m))^+ < \infty$ for all $m \in I$, $m \leq n_0$. The inequality (2.2) with $n$ replacing $n_0$ completes the proof.

**Examples**

If $\{x_n, \mathscr{F}_n, n \in I\}$ is a martingale, then $\{|x_n|, \mathscr{F}_n, n \in I\}$ is a submartingale as is $\{x_n^2, \mathscr{F}_n, n \in I, n \leq n_0\}$ for any $n_0$ such that $Ex_{n_0}^2 < \infty$. If $\{x_n, \mathscr{F}_n, n \in I\}$ is a submartingale, then

$$\{\max(x_n, a), \mathscr{F}_n, n \in I\}$$

is a submartingale for any real $a$; in particular $\{x_n^+, \mathscr{F}_n, n \in I\}$ is a submartingale.

**Lemma 2.2** Suppose that $\{x_n, \mathscr{F}_n, n \in I\}$ is a submartingale and that $I$ contains its supremum. Then for every real $a$ $\{\max(x_n, a), \mathscr{F}_n, n \in I\}$ is a uniformly integrable submartingale. If in addition $\{x_n, \mathscr{F}_n, n \in I\}$ is a martingale, then it is uniformly integrable.

***Proof*** Let $m$ denote the supremum of $I$, and put $x_n(a) = \max(x_n, a)$. By Lemma 2.1 $\{x_n(a), \mathscr{F}_n, n \in I\}$ is a submartingale. Hence for any $c > 0$, $n \in I$,

$$\begin{aligned} cP\{x_n(a) > c\} &\leq \int_{(x_n(a) > c)} x_n(a) \\ &\leq \int_{(x_n(a) > c)} x_m(a) \leq E[x_m^+(a)]. \end{aligned} \tag{2.3}$$

The extreme terms of (2.3) show that $P\{x_n(a) > c\} \to 0$ uniformly in $n$ as $c \to \infty$. The intermediate terms then establish the desired uniform integrability. If $\{x_n, \mathscr{F}_n, n \in I\}$ is a martingale, it is uniformly integrable by an application of the previous result to the submartingale $\{|x_n|, \mathscr{F}_n, n \in I\}$.

Let $\{x_n, \mathscr{F}_n, 0 \leq n < \infty\}$ be a submartingale with $Ex_0 > -\infty$. Put $y_0 = x_0$, $y_n = x_n - x_{n-1}$ $(n > 0)$. Then $x_n = \sum_{k=0}^{n} y_k$. Let $u_0 = 1$ and for each $k > 0$ let $u_k$ be an $\mathscr{F}_{k-1}$-measurable r.v. taking on the values 0 and 1. Set $\hat{x}_n = \sum_{k=0}^{n} u_k y_k$ $(0 \leq n < \infty)$. Clearly $E\hat{x}_n$ exists, and for any $n = 1, \ldots$

$$(2.4) \qquad E\hat{x}_n = Ey_0 + \sum_{k=1}^{n} \int_{\{u_k=1\}} y_k = Ey_0 + \sum_{k=1}^{n} \int_{\{u_k=1\}} E(y_k \mid \mathscr{F}_{k-1}) \leq Ey_0 + \sum_{k=1}^{n} Ey_k = Ex_n.$$

The inequality (2.4) may be given the following interpretation. Let $x_0$ be the initial fortune of a gambler and $y_1, y_2, \ldots$ successive increments to that fortune obtained by playing a sequence of "favorable" $(E(y_k \mid \mathscr{F}_{k-1}) \geq 0)$ games of chance. The sequence $(u_k)$ is the gambler's system for skipping certain games in the sequence. The requirement that $u_k$ be $\mathscr{F}_{k-1}$-measurable states that the gambler's decision to participate in the game indexed by $k$ must be a function of his past experience only. The inequality (2.4) then states that at the end of the game indexed by $n$ the expected fortune of a gambler using such a system is majorized by that of a gambler who plays every time.

We are now in a position to prove the so-called upcrossing inequality, which is used in proving the martingale convergence theorem. Let $a_1, a_2, \ldots, a_n$ be any real numbers, and let $(r, s)$ be any non-empty interval. The number of *upcrossings* of $(r, s)$ by $a_1, \ldots, a_n$ is by definition the number of times that the values $a_k$, $1 \leq k \leq n$ pass from being $\leq r$ to being $\geq s$. Formally, let $t_0 = 0$, and for each $m = 1, 2, \ldots$ let $t_m$ be the least $k > t_{m-1}$ (if any) for which

$$a_k \leq r \quad (m \text{ odd})$$

$$a_k \geq s \quad (m \text{ even}).$$

Then the number of upcrossings is $\beta$, where $2\beta$ is the largest even index $m$ for which $t_m$ is defined.

Now suppose that $\{x_i, \mathscr{F}_i, 1 \leq i \leq n\}$ is a non-negative submartingale, and take $a_i = x_i$ and $r = 0$ in the preceding paragraph. Put $x_0 \equiv 0$, $u_0 \equiv 1$ and define $u_1, \ldots, u_n$ as follows:

$$u_i = 1 \quad \text{if } t_m < i \leq t_{m+1} \qquad (m \text{ odd})$$

$$= 0 \quad \text{if } t_m < i \leq t_{m+1} \qquad (m \text{ even}).$$

where $t_1 = n$ if $t_1$ is not otherwise defined. Then it is clear that

$$s\beta \leq \sum_{i=0}^{n} u_i y_i = \hat{x}_n,$$

where as above we have put $y_i = x_i - x_{i-1}$ $(1 \leq i \leq n)$. Since for each $i = 1, \ldots, n$

$$\{u_i = 1\} = \bigcup_{m \text{ odd}} (\{t_m < i\} - \{t_{m+1} < i\}) \in \mathscr{F}_{i-1},$$

it follows from (2.4) that

$$sE\beta \leq E\hat{x}_n \leq Ex_n.$$

By observing that the number of upcrossings of an arbitrary interval $(r, s)$ by a submartingale $\{x_i, \mathscr{F}_i, 1 \leq i \leq n\}$ is the same as the number of upcrossings of $(0, s - r)$ by the non-negative submartingale $\{(x_i - r)^+, \mathscr{F}_i, 1 \leq i \leq n\}$ we have proved the following:

**Lemma 2.3 Upcrossing Inequality** Let $\{x_i, \mathscr{F}_i, 1 \leq i \leq n\}$ be a submartingale. The number of upcrossings $\beta$ of an arbitrary nonempty interval $(r, s)$ by $x_1, x_2, \ldots, x_n$ satisfies

$$E\beta \leq (s - r)^{-1}E(x_n - r)^+ \leq (s - r)^{-1}[Ex_n^+ + |r|].$$

**Theorem 2.1 Martingale Convergence Theorem** Let $\{x_n, \mathscr{F}_n, -\infty < n < \infty\}$ be a submartingale. Set $\mathscr{F}_{-\infty} = \bigcap_{-\infty}^{0} \mathscr{F}_n$, $\mathscr{F}_\infty = \mathscr{B}\left(\bigcup_{1}^{\infty} \mathscr{F}_n\right)$. Then

(a) $$x_{-\infty} = \lim_{n \to -\infty} x_n \text{ exists a.s.,}$$

(b) $$Ex_{-\infty}^+ < \infty,$$

(c) $$\{x_n, \mathscr{F}_n, -\infty \leq n < \infty\} \quad \text{is a submartingale.}$$

If $\sup_n Ex_n^+ < \infty$, then

(d) $$x_\infty = \lim_{n \to \infty} x_n \text{ exists a.s.,}$$

(e) $Ex_\infty^+ < \infty$ and if for some $n \geq -\infty$

$$Ex_n > -\infty, \quad \text{then} \quad E|x_\infty| < \infty,$$

(f) $\{x_n, \mathscr{F}_n, -\infty \leq n \leq \infty\}$ is a submartingale if and only if $(x_n^+)$ is uniformly integrable.

***Proof*** (a) Let $x^* = \limsup_{n\to-\infty} x_n$, $x_* = \liminf_{n\to-\infty} x_n$, and suppose that $P\{x^* > x_*\} > 0$. Since $\{x^* > x_*\}$ is the union over rational $r < s$ of the events

$$B(r, s) = \{x^* > s > r > x_*\},$$

it follows that for some rationals $r < s$, $P(B(r, s)) > 0$. Now on $B(r, s)$ the number $\beta_n$ of upcrossings of the interval $(r, s)$ by $x_{-n}, \ldots, x_0$ tends monotonically to $+\infty$ with $n$, and hence $\lim_{n\to\infty} E\beta_n = +\infty$. But by Lemma 2.3

$$E(\beta_n) \le (s - r)^{-1} E(x_0 - r)^+ < \infty. \tag{2.5}$$

This contradiction proves (a).

(b) By Lemma 2.1 $\{x_n^+, \mathscr{F}_n, -\infty < n < \infty\}$ is a submartingale, and thus $Ex_n^+$ is increasing in $n$. Hence by (a) and Fatou's lemma

$$Ex_{-\infty}^+ \le \lim_{n\to-\infty} Ex_n^+ < Ex_0^+ < \infty.$$

(c) Let $x_n(a) = \max(x_n, a)(-\infty \le n < \infty)$. By Lemma 2.2 $\{x_n(a), \mathscr{F}_n, -\infty < n \le 0\}$ is a uniformly integrable submartingale. Let $-\infty < m < n < \infty$. Then for any $A \in \mathscr{F}_{-\infty} \subset \mathscr{F}_m$

$$\int_A x_m(a) \le \int_A x_n(a).$$

Letting $m \to -\infty$ we have by uniform integrability and (a)

$$\int_A x_{-\infty}(a) \le \int_A x_n(a).$$

Letting $a \to -\infty$ proves (c).

(d) An argument similar to that of (a) with the right-hand side of (2.5) replaced by $(s - r)^{-1} E(x_n^+ + |r|)$, which remains bounded as $n \to \infty$ by assumption, proves (d).

(e) By Fatou's lemma, $Ex_\infty^+ \le \sup_n Ex_n^+ < \infty$. To prove the second part of (e) we note that $E|x_n| = Ex_n^+ + Ex_n^- = 2Ex_n^+ - Ex_n$. Let $n_0$ be an integer for which $Ex_{n_0} > -\infty$. Since $Ex_n$ is increasing, we have by Fatou's lemma

$$E|x_\infty| \le \sup_{n\ge n_0} E|x_n| = \sup_{n\ge n_0} (2Ex_n^+ - Ex_n)$$

$$\le 2 \sup_{n\ge n_0} Ex_n^+ - Ex_{n_0} < \infty.$$

(f) The proof of the "if" part of (f) is similar to that of (c). Suppose then that $\{x_n, \mathscr{F}_n, -\infty \le n \le \infty\}$ and hence by Lemma 2.1 that $\{x_n^+, \mathscr{F}_n, -\infty \le n \le \infty\}$ is a submartingale. Lemma 2.2 shows that $(x_n^+)$ is uniformly integrable.

## 2. Applications of the Martingale Convergence Theorem

### a. P. Lévy's Theorem (Theorem 1.4)

Let $\mathscr{F}_1 \subset \mathscr{F}_2 \subset \cdots \subset \mathscr{F}$, $\mathscr{F}_\infty = \mathscr{B}\left(\bigcup_1^\infty \mathscr{F}_m\right)$, and let $z$ be a non-negative $\mathscr{F}$-measurable r.v. with finite expectation. Then $\{E(z \mid \mathscr{F}_n), \mathscr{F}_n, 1 \le n \le \infty\}$ is a martingale (Example (a) of Section 2.1) which by Lemma 2.2 is uniformly integrable. By Theorem 2.1 $x_\infty = \lim_{n\to\infty} E(z \mid \mathscr{F}_n)$ exists with probability one, and

$$(2.6) \qquad \int_A x_\infty = \int_A E(z \mid \mathscr{F}_n) = \int_A z \qquad (A \in \mathscr{F}_n, n = 1, 2, \ldots).$$

Thus (2.6) holds for all $A \in \bigcup_1^\infty \mathscr{F}_n$, and since the extreme terms of (2.6) are finite measures, i.e., probability measures except for (1.4), on $\mathscr{F}_\infty$ it follows by the uniqueness part of the extension theorem (Theorem 1.1) that (2.6) holds for all $A \in \mathscr{F}_\infty$. Hence $x_\infty = E(z \mid \mathscr{F}_\infty)$ by the definition of conditional expectation, and thus $\lim_{n\to\infty} E(z \mid \mathscr{F}_n) = E(z \mid \mathscr{F}_\infty)$.

### b. Kolmogorov's 0 – 1 Law

Let $y_1, y_2, \ldots$ be independent random variables, $\mathscr{F}_n = \mathscr{B}(y_1, \ldots, y_n)$, and let $A$ be a "tail event," i.e., an event such that for every $n = 1, 2, \ldots, A \in \mathscr{B}(y_{n+1}, y_{n+2}, \ldots)$. Then $P(A \mid \mathscr{F}_n) = P(A)$ for every $n$. But by **a.** $P(A \mid \mathscr{F}_n) \to I_A$, and it follows that $P(A)$ must be 0 or 1.

### c. Strong Law of Large Numbers

With the assumptions of Example (d) of Section 2.1, we see by part (a) of Theorem 2.1 and **b.** above that

$$x_{-\infty} = \lim_{n\to\infty} s_n/n$$

exists and is constant a.s. Since by Theorem 2.1(c) $Ex_{-\infty} = Ex_{-1} = Ey_1$, it follows that $s_n/n \to Ey_1$.

### d. Likelihood Ratios

Suppose that $\{\mathscr{F}_n, n \geq 1\}$, $Q$, and $\{x_n, n \geq 1\}$ are as in Example (c) of Section 1. If $Q$ is absolutely continuous with respect to $P$ on $\mathscr{F}_\infty$, i.e., if there exists a r.v. $x$ for which

$$Q(A) = \int_A x \quad \text{for every} \quad A \in \mathscr{F}_\infty,$$

then $x_n = E(x \mid \mathscr{F}_n)$, and **a.** above applies. Suppose, on the other hand, that $Q$ is singular, i.e., that there exists an $S \in \mathscr{F}_\infty$ for which $Q(S) = 0$ and $P(S) = 1$. Then from $Q(A) = \int_A x_n$ ($A \in \mathscr{F}_n$; $n = 1, 2, \ldots$), Theorem 2.1, and Fatou's lemma we have

$$\int_A x_\infty \leq Q(A) \quad \text{for all} \quad A \in \bigcup_1^\infty \mathscr{F}_n. \tag{2.7}$$

By the uniqueness part of the extension theorem, (2.7) holds for all $A \in \mathscr{B}\left(\bigcup_1^\infty \mathscr{F}_n\right) = \mathscr{F}_\infty$. Putting $A = S$, we have $P\{x_\infty = 0\} = 1$.

In general $Q = pQ_1 + (1 - p)Q_2 (0 \leq p \leq 1)$, where $Q_1$ is absolutely continuous and $Q_2$ singular. If $0 < p < 1$, then $0 < P\{x_\infty = 0\} < 1$. Since, as is easily seen, $\{x_n^{1/2}, \mathscr{F}_n, 1 \leq n < \infty\}$ is uniformly integrable, it follows that $P\{x_\infty = 0\} = 1$, i.e., $Q$ is singular, if and only if $\lim_{n \to \infty} Ex_n^{1/2} = 0$. In the case that $y_1, y_2, \ldots$ are independent with probability density function $f$ with respect to some $\sigma$-finite measure $\mu$ and $g$ is some other density with respect to $\mu$, $\mathscr{F}_n = \mathscr{B}(y_1, \ldots, y_n)$, and $x_n = \prod_1^n \frac{g(y_k)}{f(y_k)}$ $(n \geq 1)$, then by the Cauchy-Schwarz inequality

$$E\left(\frac{g(y_k)}{f(y_k)}\right)^{1/2} = \int (gf)^{1/2}\, d\mu < \left[\int g\, d\mu \int f\, d\mu\right]^{1/2} = 1.$$

Hence $Ex_n^{1/2} = \left[E\left(\frac{g(y_1)}{f(y_1)}\right)^{1/2}\right]^n \to 0$ $(n \to \infty)$, and it follows that $P\{x_n \to 0\} = 1$.

## 3. Stopping Times-Definition and Fundamental Properties

Let $(\Omega, \mathscr{F}, P)$ be a probability space and $\{\mathscr{F}_n, n \in I\}$ an increasing family of sub-$\sigma$-algebras of $\mathscr{F}$. A *stopping time* is by definition a random variable $t$ such that

$$P\{t \in I\} = 1, \tag{i}$$

and

(ii) $\{t = n\} \in \mathscr{F}_n$ for each $n \in I$.

If for each $n \in I$, $x_n$ is a random variable which is $\mathscr{F}_n$-measurable and $t$ is a stopping time, then

$$x_t \equiv \sum_{n \in I} x_n I\{t = n\}$$

is a random variable, since for any Borel set $B$

$$\{x_t \in B\} = \bigcup_{n \in I} \{x_n \in B, t = n\} \in \mathscr{F}.$$

The collection of all sets $A \in \mathscr{F}$ such that $A \cap \{t = n\} \in \mathscr{F}_n$ for all $n \in I$ is a sub-$\sigma$-algebra of $\mathscr{F}$, which will be denoted by $\mathscr{F}_t$. It is easy to see that $t$ and $x_t$ are $\mathscr{F}_t$-measurable.

If we interpret the values $x_1, x_2, \ldots$ of a martingale $\{x_n, \mathscr{F}_n, 1 \leq n < \infty\}$ as the values assumed by a gambler's fortune as he plays a sequence of fair games, then a stopping time is a strategy by which the gambler decides when to stop playing, and $x_t$ is his terminal fortune. The requirement $\{t = n\} \in \mathscr{F}_n$ states that the gambler's decision whether to stop at time $n$ may depend on his past experience but not on the yet unobserved future. It is natural to inquire whether

$$Ex_t = Ex_1,$$

i.e., whether the property of "fairness" is preserved under any stopping time $t$. This section and those following are devoted to exploring this question. We note that the answer cannot be an unqualified affirmative, for if $x_n$ is the difference between the number of heads and the number of tails in $n$ tosses of a fair coin and if $t =$ first $n \geq 1$ such that $x_n = 1$ ($= \infty$ if no such $n$ exists), then $P\{t < \infty\} = 1$ and $Ex_t = 1 \neq 0 = Ex_1$. However,

**Theorem 2.2** Let $\{x_n, \mathscr{F}_n, n \in I\}$ be a submartingale and $t$ a stopping time. Let $t(n) = \min(t, n)$ $(n \in I)$.
(a) If for some $N \in I$, $P\{t \leq N\} = 1$, then $x_t \leq E(x_N \mid \mathscr{F}_t)$.
(b) $\{x_{t(n)}, \mathscr{F}_n, n \in I\}$ is a submartingale. In particular if $+\infty \in I$, then for each $n \in I$

(2.8) $$E(x_t \mid \mathscr{F}_n) \geq x_n \quad \text{on} \quad \{t \geq n\}.$$

(c) If $P\{t < \infty\} = 1$ and $Ex_t$ exists, and if

$$\liminf_{n\to\infty} \int_{\{t>n\}} x_n^+ = 0, \tag{2.9}$$

then (2.8) holds for each $n \in I$.

***Proof*** (a) By Lemma 2.1 for all $n \in I$, $n \leq N$, $A \in \mathscr{F}_t$,

$$\int_{A(t=n)} x_n^+ \leq \int_{A(t=n)} x_N^+. \tag{2.10}$$

Summing (2.10) on $n$ ($n \in I$, $n \leq N$) shows that

$$Ex_t^+ \leq Ex_N^+ < \infty; \tag{2.11}$$

and the same argument with $x_n(x_N)$ in place of $x_n^+(x_N^+)$ proves (a).
(b) By (2.11) $Ex_{t(n)}^+ < \infty$ for all $n \in I$. Let $n \in I$, $-\infty < n < \infty$. Then for any $A \in \mathscr{F}_n$

$$\int_A x_{t(n)} = \int_{A(t\leq n)} x_t + \int_{A(t>n)} x_n \leq \int_{A(t\leq n)} x_t + \int_{A(t>n)} E(x_{n+1} \mid \mathscr{F}_n)$$

$$= \int_{A(t\leq n+1)} x_t + \int_{A(t>n+1)} x_{n+1} = \int_A x_{t(n+1)}.$$

Hence $\{x_{t(n)}, \mathscr{F}_n, n \in I, -\infty < n < \infty)$ is a submartingale. Assume now that $+\infty \in I$ and, moreover, that $\mathscr{F}_\infty = \mathscr{B}\left(\bigcup_1^\infty \mathscr{F}_n\right)$, $x_\infty = \lim x_n$. By (a) $Ex_t^+ \leq Ex_\infty^+ < \infty$, and by Lemma 2.2 $(x_n^+)$ is uniformly integrable. Since $x_{t(n)}^+ \leq x_t^+ + x_n^+$ it follows that $(x_{t(n)}^+)$ is uniformly integrable, and hence by Theorem 2.1 (f) that $\{x_{t(n)}, \mathscr{F}_n, n \in I, -\infty < n \leq \infty\}$ is a submartingale. In general $\mathscr{F}_\infty \supset \mathscr{B}\left(\bigcup_1^\infty \mathscr{F}_n\right)$ and a short additional argument is required; we omit the details. The case $-\infty \in I$ is treated similarly.
(c) Let $n \in I - \{+\infty\}$ and $A \in \mathscr{F}_n$. By (b) for each integer $m > n$

$$\int_{A(t\geq n)} x_n \leq \int_{A(t\geq n)} x_{t(m)} \leq \int_{A(n\leq t\leq m)} x_t + \int_{A(t>m)} x_m^+.$$

Let $m \to \infty$ along a subsequence $m'$ for which

$$\int_{\{t>m'\}} x_{m'}^+ \to \liminf_m \int_{\{t>m\}} x_m^+ = 0.$$

Since $P\{t < \infty\} = 1$, $\int_{A\{n \le t \le m'\}} x_t \to \int_{A\{t \ge n\}} x_t$ and hence

$$\int_{A(t \ge n)} x_n \le \int_{A(t \ge n)} x_t.$$

This completes the proof.

The second of the following two lemmas is sometimes helpful in trying to prove that (2.8) holds. Let $\{x_n, \mathscr{F}_n, 1 \le n < \infty\}$ be a submartingale with $Ex_1 \ge 0$, and put $\mathscr{F}_0 = \{\phi, \Omega\}$, $x_0 = 0$, $y_n = x_n - x_{n-1}$, $a_n = E(|y_n| \mid \mathscr{F}_{n-1})$, $b_n = E(y_n^+ \mid \mathscr{F}_{n-1})$, $\sigma_n^2 = E(y_n^2 \mid \mathscr{F}_{n-1})$ $(n = 1, 2, \ldots)$.

**Lemma 2.4** If $y_n \ge 0$ $(n = 1, 2, \ldots)$, then for any stopping time $t$

$$E\left(\sum_1^t y_k\right) = E\left(\sum_1^t E(y_k \mid \mathscr{F}_{k-1})\right).$$

***Proof***

$$E\left(\sum_1^t y_k\right) = E\left(\sum_1^\infty I_{\{t \ge k\}}\, y_k\right) = \sum_1^\infty \int_{\{t \ge k\}} y_k$$

$$= \sum_1^\infty \int_{\{t \ge k\}} E(y_k \mid \mathscr{F}_{k-1}) = E\left(\sum_1^t E(y_k \mid \mathscr{F}_{k-1})\right),$$

where the last equality follows by reversing the steps which led to the first two.

**Lemma 2.5** If $t$ is any stopping time for which

$$E\left(\sum_1^t b_k\right) < \infty, \tag{2.12}$$

then (2.8) holds $(n = 1, 2, \ldots)$.

***Proof*** By Theorem 2.2(c) it suffices to verify that $Ex_t$ exists and that (2.9) holds. By (2.12) and Lemma 2.4 $E\left(\sum_1^t y_k^+\right) = E\left(\sum_1^t b_k\right) < \infty$. Hence $Ex_t^+ \le E\left(\sum_1^t y_k^+\right) < \infty$, and

$$\int_{\{t > n\}} x_n^+ \le \int_{\{t > n\}} \left(\sum_1^n y_k^+\right) \le \int_{\{t > n\}} \left(\sum_1^t y_k^+\right) \to 0 \ (n \to \infty).$$

**Theorem 2.3** Suppose that $\{x_n, \mathscr{F}_n, 1 \leq n < \infty\}$ is a martingale with $Ex_1 = 0$ and $t$ is any stopping time.
(a) If

$$\text{(2.13)} \qquad Ex_t \text{ exists} \quad \text{and} \quad \liminf \int_{\{t>n\}} |x_n| = 0,$$

then $Ex_t = 0$; more generally

$$E(x_t \mid \mathscr{F}_n) = x_n \qquad \text{on } \{t \geq n\}, n = 1, 2, \ldots .$$

(b) If $Ey_n^2 < \infty, n = 1, 2, \ldots,$ and the second part of (2.13) holds, then $Ex_t^2 = E\left(\sum_1^t \sigma_n^2\right)$.

(c) In order that (2.13) hold, it suffices that either $E\left(\sum_1^t a_n\right) < \infty$ or $E\left(\sum_1^t \sigma_n^2\right) < \infty$.

***Proof*** (a) and the first part of (c) follow at once from Theorem 2.2(c) and the proof of Lemma 2.5.
(b) Let $z_n = x_n^2 - \sum_1^n \sigma_n^2$. It is easy to see that $\{z_n, \mathscr{F}_n, 1 \leq n < \infty\}$ is a martingale. For $n = 1, 2, \ldots$ let $t(n) = \min(t, n)$. Then by (a)

$$Ez_{t(n)} = 0$$

and hence

$$Ex_{t(n)}^2 = E\left(\sum_1^{t(n)} \sigma_k^2\right).$$

As $n \to \infty$, $t(n) \to t$; and by Fatou's lemma and the monotone convergence theorem

$$\text{(2.14)} \qquad Ex_t^2 \leq \lim Ex_{t(n)}^2 = \lim E\left(\sum_1^{t(n)} \sigma_k^2\right) = E\left(\sum_1^t \sigma_k^2\right).$$

For the reverse inequality it suffices to assume $Ex_t^2 < \infty$ and show that

$$Ex_{t(n)}^2 \leq Ex_t^2, \qquad n = 1, 2, \ldots .$$

But for each $n$, by Theorem 2.2(c) and Lemma 2.2

$$E(|x_t| \mid \mathscr{F}_n) \geq |x_n| \quad \text{on} \quad \{t \geq n\},$$

and hence by the Schwarz inequality

$$E(x_t^2 \mid \mathscr{F}_n) \geq (E(|x_t| \mid \mathscr{F}_n))^2 \geq x_n^2 \quad \text{on} \quad \{t \geq n\}.$$

It follows that

$$Ex_t^2 = \int_{\{t \le n\}} x_t^2 + \int_{\{t > n\}} x_t^2 \ge \int_{\{t \le n\}} x_t^2 + \int_{\{t > n\}} x_n^2 = Ex_{t(n)}^2.$$

(c) Assume that $E\left(\sum_1^t \sigma_n^2\right) < \infty$ and also that $Ey_n^2 < \infty, n \ge 1$.

From (2.14) and the Schwarz inequality

$$E|x_t| \le (Ex_t^2)^{1/2} < \infty \quad \text{and} \quad B = \liminf \int_{\{t > n\}} x_n^2 < \infty.$$

Thus

$$\liminf \int_{\{t > n\}} |x_n| = \liminf \int_{\{t > n, |x_n| > a\}} |x_n|$$

$$\le \frac{1}{a} \liminf \int_{\{t > n, |x_n| > a\}} x_n^2 \le Ba^{-1} \to 0 \quad \text{as} \quad a \to \infty.$$

By putting $y_n' = y_n I_{\{t \ge n\}}$ the assumption that $Ey_n^2 < \infty$ can now be dropped.

If the $y$'s are independent with mean 0, the r.v.'s $a_n$ and $\sigma_n^2$ are constants. In this special case the conditions of Theorem 2.3(c) are frequently quite easy to verify. For example, if the $y$'s have uniformly bounded absolute first moments, then $Ex_t = 0$ for all $t$ such that $Et < \infty$. (If the $y$'s are identically distributed, this result is known as Wald's lemma.) If $\sigma_1^2 = \sigma_2^2 = \cdots = \sigma^2 < \infty$, then $Et < \infty$ implies that $Ex_t = 0$ and $Ex_t^2 = \sigma^2 Et$. (See Theorems 2.4 and 2.5 for further applications of Theorem 2.3 in the independent case.)

## 4. Applications of Stopping Times

### a. Kolmogorov's Inequalities

Let $n$ be a positive integer, and let $\{x_k, \mathscr{F}_k, 1 \le k \le n\}$ be a non-negative submartingale. Then for any $\varepsilon > 0$

$$\varepsilon P\{\max_{1 \le k \le n} x_k > \varepsilon\} \le \int_{\{\max_{1 \le k \le n} x_k > \varepsilon\}} x_n \le Ex_n. \tag{2.16}$$

In fact, if $A = \{\max_{1 \le k \le n} x_k > \varepsilon\}$ and

$$t = \text{first } k \ge 1 \quad \text{such that} \quad x_k > \varepsilon \quad \text{on} \quad A$$
$$= n \text{ off } A,$$

then it is easily seen that $A \in \mathcal{F}_t$ and hence by Theorem 2.2(a)

$$\varepsilon P(A) \leq \int_A x_t \leq \int_A x_n \leq Ex_n.$$

If $x_k = s_k^2$, where $s_k$ is the $k$th partial sum of a sequence of independent r.r.v.'s with mean 0 and finite variances, this result gives one of the Kolmogorov inequalities. Another is as follows.

Let $y_1, \ldots, y_n$ be independent with $Ey_k = 0$, $Ey_k^2 = \sigma_k^2$ $(1 \leq k \leq n)$. Let $z_n = \max_{1 \leq k \leq n} |y_k|$, $x_k = \sum_1^k y_i$ $(1 \leq k \leq n)$. Then for any $\varepsilon > 0$

$$P\{\max_{1 \leq k \leq n} |x_k| > \varepsilon\} \geq 1 - \frac{E(\varepsilon + z_n)^2}{\sum_1^n \sigma_k^2}. \tag{2.17}$$

***Proof*** Let $\mathcal{F}_k = (y_1, \ldots, y_k)$ $(1 \leq k \leq n)$, and put $A = \{\max_{1 \leq k \leq n} |x_k| > \varepsilon\}$. Let

$$\begin{aligned} t &= \text{first } k \geq 1 \quad \text{such that} \quad |x_k| > \varepsilon \quad \text{on} \quad A \\ &= n \text{ off } A. \end{aligned}$$

Then $t$ is a stopping time; and by Theorem 2.3

$$\left(\sum_1^n \sigma_k^2\right) P(\Omega - A) \leq E\left(\sum_1^t \sigma_k^2\right) = Ex_t^2 \leq E(\varepsilon + z_n)^2,$$

which is equivalent to (2.17).

### b. Hájek-Rényi-Chow Inequality

The following result generalizes the first Kolmogorov inequality (2.16). Let $\{x_k, \mathcal{F}_k, 0 \leq k \leq n\}$ be a non-negative submartingale with $x_0 = 0$, and let $(c_k)$ be a sequence of non-negative constants. Then

$$c_k x_k = \sum_{i=1}^k (c_i x_i - c_{i-1} x_{i-1}) \leq \sum_{i=1}^k [c_i(x_i - x_{i-1}) + (c_i - c_{i-1})^+ x_{i-1}].$$

Obviously

$$\left\{\sum_{i=1}^k [c_i(x_i - x_{i-1}) + (c_i - c_{i-1})^+ x_{i-1}], \mathcal{F}_k, 1 \leq k \leq n\right\}$$

is a non-negative submartingale, and hence by (2.16)

$$P\{\max_{1 \leq k \leq n} c_k x_k \geq 1\} \leq \sum_{k=1}^n [c_k E(x_k - x_{k-1}) + (c_k - c_{k-1})^+ Ex_{k-1}].$$

## c. Ballot Theorems

Let $y_1, y_2, \ldots, y_n$ be independent, identically distributed (or more generally, exchangeable), non-negative, integer-valued random variables with finite expectation, and put $s_k = \sum_1^k y_i$ $(1 \leq k \leq n)$. Then

$$(2.18)\quad P\{s_k < k \quad \text{for all} \quad 1 \leq k \leq n \mid s_n\} = \left(1 - \frac{s_n}{n}\right)^+.$$

***Proof*** Let $\mathscr{F}_{-k} = \mathscr{B}(s_k, s_{k+1}, \ldots, s_n)$, $x_{-k} = \dfrac{s_k}{k}$ $(1 \leq k \leq n)$. $\{x_k, \mathscr{F}_k, -n \leq k \leq -1\}$ is a martingale (see Example (d) Section 2.1). Since (2.18) is trivially satisfied on $\{s_n \geq n\}$, assume $s_n < n$. Define

$$t = \inf\{k: -n \leq k \leq -1, x_k \geq 1\},$$

where the inf of the empty set is $-1$. Since $x_t = 1$ on $\left\{\max_{1 \leq k \leq n} \dfrac{s_k}{k} \geq 1\right\}$ and 0 elsewhere, we have by Theorem 2.2, on $\{s_n < n\}$

$$P\{s_k \geq k \text{ for some } 1 \leq k \leq n \mid s_n\} = E(x_t \mid \mathscr{F}_{-n}) = x_{-n} = \frac{s_n}{n},$$

which implies (2.18).

## d. Borel-Cantelli-Lévy Lemma

Let $B_1, B_2, \ldots$ be a sequence of events and $\{\mathscr{F}_n, 0 \leq n < \infty\}$ a non-decreasing sequence of sub-$\sigma$-algebras of $\mathscr{F}$ such that $B_n \in \mathscr{F}_n$ $(1 \leq n < \infty)$. Then except for an event of probability 0

$$\sum_1^\infty I_{B_n} < \infty \quad \text{if and only if} \quad \sum_1^\infty P(B_n \mid \mathscr{F}_{n-1}) < \infty.$$

***Proof*** Let $x_n = \sum_1^n [I_{B_k} - P(B_k \mid \mathscr{F}_{k-1})]$ $(1 \leq n < \infty)$. Then $\{x_n, \mathscr{F}_n, 1 \leq n < \infty\}$ is a martingale. For any $c > 0$ let

$$\begin{aligned} t &= \text{first } n \geq 1 \quad \text{such that} \quad x_n \geq c \\ &= \infty \quad \text{if} \quad \text{no such } n \text{ exists,} \end{aligned}$$

and put $t(n) = \min(t, n)$. By Theorem 2.2(b) $\{x_{t(n)}, \mathscr{F}_n, 1 \leq n < \infty\}$ is a martingale, and clearly $x_{t(n)} \leq c + 1$ for all $n = 1, 2, \ldots$. Hence by Theorem 2.1 $\lim_{n\to\infty} x_{t(n)}$ exists and is finite with probability one, i.e., $\lim_{n\to\infty} x_n$ exists and is finite on $\{t = \infty\}$. Since

$\{\sup x_n < \infty\} = \bigcup_{c=1}^{\infty} \{\sup x_n < c\}$, we see that $\lim_{n\to\infty} x_n$ exists and is finite on $\{\sup x_n < \infty\}$, and hence that

$$\sum_1^\infty P(B_k \mid \mathscr{F}_{k-1}) < \infty \quad \text{on} \quad \left\{\sum_1^\infty I_{B_k} < \infty\right\}.$$

The same argument applied to $\{-x_n, \mathscr{F}_n, 1 \leq n < \infty\}$ completes the proof.

## e. Chung-Fuchs Theorem

Let $y_1, y_2, \ldots,$ be i.i.d. with $Ey_1 = 0$, $P\{y_1 = 0\} < 1$, and let $s_n = \sum_1^n y_k$. A theorem of Chung and Fuchs [1] implies that $P\{\limsup_n s_n = +\infty, \liminf_n s_n = -\infty\} = 1$. To prove this result let $q = P\{s_n < 0 \text{ for all } n\}$ and for each $n = 1, 2, \ldots, A_n = \{s_k < s_n \text{ for all } k \neq n\}$. Then

$$\begin{aligned} P(A_n) &= P\{s_n - s_k > 0 \text{ for all } k < n \text{ and } s_k - s_n < 0 \text{ for all } k > n\} \\ &= P\{s_n - s_k > 0 \text{ for all } k < n\}q. \end{aligned}$$

Let

$$\begin{aligned} t &= \text{first } n \geq 1 \quad \text{such that} \quad s_n \leq 0 \\ &= \infty \quad \text{if} \quad \text{no such } n \text{ occurs.} \end{aligned}$$

Then $P\{s_n - s_k > 0 \text{ for all } k < n\} = P\{s_i > 0 \text{ for all } 1 \leq i \leq n - 1\} = P(t \geq n)$. Hence $P(A_n) = qP(t \geq n)$ and $1 \geq \sum_1^\infty P(A_n) = q \sum_1^\infty P(t \geq n) = qEt$. If $q > 0$, then $Et < \infty$. By Wald's lemma $Es_t = 0$ and hence $P(s_t = 0) = 1$. This contradicts the hypothesis that $P(y_1 = 0) < 1$ and hence $q = 0$. Now let $t_0 = s_0 = 0$ and for each $n = 1, 2, \ldots,$

$$t_n = \inf\{k: k > t_{n-1}, s_k \geq s_{t_{n-1}}\}.$$

Then $P(t_n < \infty) = 1$ for all $n = 1, 2, \ldots, s_{t_1} - s_{t_0}, s_{t_2} - s_{t_1}, \ldots$ are i.i.d. non-negative random variables (not identically 0), and hence $P\{\lim_n s_{t_n} = +\infty\} = 1$. This proves that $\limsup_n s_n = +\infty$. The lim inf case follows by symmetry.

### f. Wald's Lemma when $Et^{1/\alpha} < \infty$.

Let $y_1, y_2, \ldots$ be independent with $Ey_k = 0$ $(k = 1, 2, \ldots)$ such that for some $1 < \alpha \leq 2$

$$\sup_n n^{-1} \sum_1^n E|y_k|^\alpha = B < \infty.$$

Let $\mathscr{F}_n = \mathscr{B}(y_1, \ldots, y_n)$ and $s_n = \sum_1^n y_k$. Then $Es_t = 0$ for each stopping time $t$ for which $Et^{1/\alpha} < \infty$. (The case $\alpha = 2$ was obtained by Burkholder and Gundy [1]. We follow their proof). By Theorem 2.2 $Es_{\min(t,n)} = 0$ and hence by the dominated convergence theorem it suffices to prove

$$E[\sup_n |s_{\min(t,n)}|] < \infty.$$

For each $u > 0$, we have by Theorem 2.2 and by (2.16)

$$\text{(2.19)} \quad P\{\sup_n |s_{\min(t,n)}| \geq u\} \leq P\{t \geq u^\alpha\} + P\{\sup_{n < u^\alpha} |s_{\min(t,n)}| \geq u\}$$
$$\leq P\{t \geq u^\alpha\} + u^{-\alpha}E[|s_{\min(t,[u^\alpha])}|^\alpha].$$

Now by an inequality of von Bahr and Esseen [1] and Lemma 2.4 we have

$$E[|s_{\min(t,[u^\alpha])}|^\alpha] \leq 2E\left(\sum_1^{\min(t,[u^\alpha])} E|y_k|^\alpha\right)$$
$$\leq 2Bu^\alpha P\{t \geq u^\alpha\} + 2B \int_{\{t < u^\alpha\}} t\, dP.$$

Hence by (2.19)

$$E[\sup_n |s_{\min(t,n)}|] = \int_0^\infty P\{\sup_n |s_{\min(t,n)}| \geq u\}\, du$$
$$\leq (1 + 2B) \int_0^\infty P\{t \geq u^\alpha\}\, du + \int_0^\infty u^{-\alpha} \int_{\{t < u^\alpha\}} t\, dP\, du$$
$$= (1 + 2B)Et^{\alpha^{-1}} + \int_\Omega t \int_{t^{\alpha^{-1}}}^\infty u^{-\alpha}\, du\, dP$$
$$= (1 + 2B)Et^{\alpha^{-1}} + (\alpha - 1)^{-1}Et^{\alpha^{-1}} < \infty.$$

## 5. Some First Passage Problems

Let $y_1, y_2, \ldots$ be independent random variables with means $\mu_1, \mu_2, \ldots$ such that for some $0 < \mu < \infty$

$$\text{(2.20)} \qquad n^{-1} \sum_1^n \mu_k \to \mu \qquad (n \to \infty).$$

Let $s_n = \sum_1^n y_k$, $\mathscr{F}_n = \mathscr{B}(y_1, \ldots, y_n)$ $(n \geq 1)$, and for each $c > 0$ define

$$t = t(c) = \text{first } n \geq 1 \quad \text{such that} \quad s_n > c$$
$$= \infty \quad \text{if} \quad \text{no such } n \text{ exists.}$$

In this section we find conditions under which as $c \to \infty$

$$Et \sim \frac{c}{\mu} \tag{2.21}$$

and

$$\operatorname{Var} t \sim (\text{const})\, c. \tag{2.22}$$

(The constant appearing in (2.22) is given in Theorem 2.5.) These results generalize well-known results of renewal theory to the effect that (2.21) holds if the $y$'s are identically distributed and non-negative and that (2.22) holds with const $= \dfrac{\sigma^2}{\mu^3}$ if in addition $\sigma^2 = Ey_1^2 - \mu^2 < \infty$.

**Theorem 2.4** If for every $\varepsilon > 0$

$$\lim_{n\to\infty} n^{-1} \sum_{k=1}^{n} \int_{\{y_k - \mu_k > \varepsilon n\}} (y_k - \mu_k) = 0, \tag{2.23}$$

and if

$$\sup_n n^{-1} \sum_1^n E(y_k - \mu_k)^- < \infty, \tag{2.24}$$

then $Et < \infty$ for every $c > 0$ and (2.21) holds as $c \to \infty$.

***Proof*** Suppose that $Et = \infty$ and put $\tau = \min(t, n)$ $(n = 1, 2, \ldots)$. Then $E\tau \to \infty$. Thus by (2.20) and Theorem 2.3

$$\mu E\tau + o(E\tau) = E\left(\sum_1^\tau \mu_k\right) = E(s_\tau) = c + E(s_\tau - c). \tag{2.25}$$

It will be shown in Lemma 2.6 that

$$E(s_\tau - c) \leq Ey_\tau^+ = o(E\tau) \qquad (n \to \infty), \tag{2.26}$$

which with (2.25) contradicts the supposition that $E\tau \to \infty$ as $n \to \infty$. Hence $Et < \infty$, and by (2.20) $E\left(\left|\sum_1^t \mu_k\right|\right) < \infty$. Since by definition $s_t > c$, it follows that $Es_t$ and $E\left(s_t - \sum_1^t \mu_k\right)$ exist. From (2.24) it follows that

$$\sup n^{-1} \sum_1^n E|y_k - \mu_k| < \infty$$

and hence that

$$E\left(\sum_1^t E|y_k - \mu_k|\right) \leq \text{const } Et < \infty.$$

Thus by (2.20) and Theorem 2.3, (2.25) holds with $\tau$ replaced by $t$. Referring once again to Lemma 2.6 below, we have $0 < E(s_t - c) \leq Ey_t = o(Et)$, which in conjunction with (2.25) (with $\tau$ replaced by $t$) proves (2.21).

The following lemma justifies (2.26) as well as the corresponding statement with $\tau$ replaced by $t$. It is stated more generally than is required, with a view to further applications (Theorem 2.5).

**Lemma 2.6** If (2.20), (2.23), and (2.24) hold, then for any increasing family $\{\tau(r), r > 0\}$ of stopping times for which

$$\infty > E\tau(r) \to \infty \qquad \text{as } r \to \infty$$

we have $Ey_\tau^+ < \infty$ for every $r > 0$ and

$$Ey_\tau^+ = o(E\tau) \qquad (r \to \infty).$$

Let $\sigma_n^2 = Ey_n^2 - \mu_n^2$, $b_n^2 = \sum_1^n \sigma_k^2$ $(n = 1, 2, \ldots)$. If

(2.27) $$\mu_n = o(n^{1/2}),$$

for some $0 < \sigma^2 < \infty$

(2.28) $$b_n^2 \sim n\sigma^2,$$

and if for every $\varepsilon > 0$

(2.29) $$\lim_{n\to\infty} n^{-1} \sum_1^n \int_{\{y_k - \mu_k > \varepsilon n^{1/2}\}} (y_k - \mu_k)^2 = 0,$$

then $E(y_\tau^+)^2 < \infty$ for every $r > 0$ and $E(y_\tau^+)^2 = o(E\tau)$ $(r \to \infty)$.

***Proof*** We shall prove the first part only; the second part is proved similarly. Let $\varepsilon > 0$ be arbitrary. Then with $\tilde{y}_k = y_k - \mu_k$, we have

$$(2.30)\quad E\tilde{y}_\tau^+ \leq \varepsilon E\tau + \int_{\{\tilde{y}_\tau > \varepsilon\tau\}} \tilde{y}_\tau \leq \varepsilon E\tau + E\left(\sum_1^\tau I_{\{\tilde{y}_k > \varepsilon k\}}\tilde{y}_k\right)$$
$$= \varepsilon E\tau + E\left(\sum_1^\tau \int_{\{\tilde{y}_k > \varepsilon k\}} \tilde{y}_k\right),$$

where the equality follows from Lemma 2.4. By (2.24) we have

$$\sup_n n^{-1} \sum_1^n E|\tilde{y}_k| = B < \infty.$$

Hence for all $n$ sufficiently large it follows from (2.23) that

$$(2.31)\quad \sum_1^n \int_{\{\tilde{y}_k > \varepsilon k\}} \tilde{y}_k \leq \sum_1^{[\varepsilon n]} E|\tilde{y}_k| + \sum_{[\varepsilon n]+1}^n \int_{\{\tilde{y}_k > \varepsilon^2 n\}} \tilde{y}_k$$
$$\leq \varepsilon B n + \varepsilon n.$$

Since by (2.20) $\mu_n = o(n)$, we have from (2.30) and (2.31)

$$Ey_\tau^+ \leq E\tilde{y}_\tau^+ + E|\mu_\tau| \leq \varepsilon(3 + B)E\tau + 0(1),$$

from which we obtain

$$Ey_\tau^+ = o(E\tau),$$

as $\varepsilon > 0$ is arbitrary.

**Theorem 2.5** Assume that

$$(2.32)\quad \sum_1^n \mu_k - n\mu = o(n^{1/2}),$$

and that (2.28) and (2.29) hold. Then as $c \to \infty$

$$(2.33)\quad Et = c/\mu + o(c^{1/2})$$

and

$$(2.34)\quad \operatorname{Var} t \sim \frac{\sigma^2}{\mu^3} c.$$

***Proof*** For ease of exposition we shall assume that $\mu_n \equiv \mu$ and $\sigma_n^2 \equiv \sigma^2$. Then as in the proof of Theorem 2.4 it may be shown that

$$(2.35)\quad Et < \infty \quad Et \sim c/\mu \qquad (c \to \infty),$$

and by using Theorem 2.3(b) it may similarly be shown that

$$Et^2 < \infty \qquad (c > 0). \tag{2.36}$$

By Lemma 2.6 and (2.35)

$$[E(s_t - c)]^2 \le E(s_t - c)^2 \le E(y_t^+)^2 = o(Et) = o(c), \tag{2.37}$$

and thus by Theorem 2.3

$$\mu Et = c + E(s_t - c) = c + o(c^{1/2}).$$

This proves (2.33).

By (2.36) and Theorem 2.3 we have

$$\begin{aligned}\frac{\sigma^2}{\mu}(c + E(s_t - c)) &= \frac{\sigma^2}{\mu} Es_t = \sigma^2 Et = E(s_t - \mu t)^2 \\ &= E(s_t - c + c - \mu t)^2 = E(s_t - c)^2 \\ &\quad + 2\mu E\left(\frac{c}{\mu} - t\right)(s_t - c) + \mu^2 E\left(t - \frac{c}{\mu}\right)^2,\end{aligned}$$

so

$$\begin{aligned}\mu^2 E\left(t - \frac{c}{\mu}\right)^2 &= \frac{\sigma^2}{\mu} c + 2\mu E\left(t - \frac{c}{\mu}\right)(s_t - c) \\ &\quad + \frac{\sigma^2}{\mu} E(s_t - c) - E(s_t - c)^2.\end{aligned} \tag{2.38}$$

By (2.37) and the Schwarz inequality

$$\begin{aligned}\left|E\left(t - \frac{c}{\mu}\right)(s_t - c)\right| &\le \left[E\left(t - \frac{c}{\mu}\right)^2 E(s_t - c)^2\right]^{1/2} \\ &= o(c^{1/2})\left[E\left(t - \frac{c}{\mu}\right)^2\right]^{1/2}.\end{aligned} \tag{2.39}$$

From (2.37)–(2.39) we have

$$\mu^2 E\left(t - \frac{c}{\mu}\right)^2 \le \frac{\sigma^2}{\mu} c + o(c^{1/2})\left[E\left(t - \frac{c}{\mu}\right)^2\right]^{1/2} + o(c)$$

and it follows that

$$E\left(t - \frac{c}{\mu}\right)^2 = 0(c). \tag{2.40}$$

Hence by (2.37)–(2.40)

$$E\left(t - \frac{c}{\mu}\right)^2 = \frac{\sigma^2}{\mu^3} c + o(c). \tag{2.41}$$

But by (2.37)

$$\begin{aligned}
\operatorname{Var} t &= E\left(t - \frac{c}{\mu}\right)^2 - \left[E\left(t - \frac{c}{\mu}\right)\right]^2 \\
&= E\left(t - \frac{c}{\mu}\right)^2 - \left[\frac{1}{\mu} E(s_t - c)\right]^2 \\
&= E\left(t - \frac{c}{\mu}\right)^2 + o(c),
\end{aligned}$$

which together with (2.41) gives (2.34).

## 6. The Martingale $x_n = dQ_n/dP_n$

In this section we shall study in somewhat more detail the class of martingales $\{x_n, \mathscr{F}_n, 1 \le n < \infty\}$ introduced in Example 2.1(c). Applications of the results are given in the following section and in the problems at the end of the chapter. By Theorem 2.1 (see also Section 2.2(d)) $x_\infty = \lim_{n\to\infty} x_n$ exists and $\{x_n, \mathscr{F}_n, 1 \le n \le \infty\}$ is a supermartingale, where as usual we have put $\mathscr{F}_\infty = \mathscr{B}\left(\bigcup_1^\infty \mathscr{F}_n\right)$. By Theorem 2.2

$$Ex_t = Ex_1 = 1$$

for any bounded stopping time $t$. It is possible to apply Theorem 2.2 to extend this equality to a larger class of stopping times, but it is easier to proceed from first principles.

For any stopping time $t$ and any $A \in \mathscr{F}_t$ (see Section 2.3 for the definition of $\mathscr{F}_t$)

$$\begin{aligned}
Q(A\{t < \infty\}) &= \sum_{n=1}^{\infty} Q(A\{t = n\}) = \sum_{n=1}^{\infty} \int_{A\{t=n\}} \frac{dQ_n}{dP_n}\, dP \\
&= \int_{A\{t<\infty\}} x_t\, dP.
\end{aligned} \tag{2.42}$$

Putting $A = \Omega$ it follows from (2.42) that

$$\int_{\{t<\infty\}} x_t\, dP = 1 \tag{2.43}$$

if and only if

$$Q\{t < \infty\} = 1. \tag{2.44}$$

It was observed in Example 2.1(c) that a special case of the martingale $\frac{dQ_n}{dP_n}$ is given by

$$\frac{dQ_n}{dP_n} = \frac{e^{\lambda s_n}}{(\varphi(\lambda))^n}, \tag{2.45}$$

where $y_1, y_2, \ldots$ are i.i.d. random variables such that $\varphi(\lambda) = E(e^{\lambda y_1}) < \infty$ for some (real) $\lambda \neq 0$, and $s_n = \sum_1^n y_k$. The following results are useful in deciding whether (2.44) (and hence (2.43)) holds for a particular stopping time $t$ when $\frac{dQ_n}{dP_n}$ is of the form (2.45).

**Lemma 2.7** Let $F(G)$ denote the distribution of $y_1$ under $P(Q)$. Then
(a) $D = \{\lambda: \varphi(\lambda) < \infty\}$ is an interval containing 0.
(b) On $D$, $\varphi$ may be differentiated any number of times (with the obvious conventions about one-sided and infinite derivatives at the end points of $D$) and

$$\varphi^{(k)}(\lambda) = \int_{-\infty}^{\infty} x^k e^{\lambda x}\, dF(x);$$

hence $\varphi^{(k)}(\lambda)/\varphi(\lambda) = \int_{-\infty}^{\infty} x^k\, dG(x)$.
(c) Unless $F$ is concentrated at 0, $\varphi$ is strictly convex on $D$.
(d) There exists at most one non-zero value $\lambda_1$ for which $\varphi(\lambda_1) = 1$. If such a value $\lambda_1$ exists, it is opposite in sign to $Ey_1$. A sufficient condition for the existence of $\lambda_1$ is that $Ey_1 \neq 0$ and that $D$ be an open interval.

***Proof*** The proof is a straightforward consequence of the convexity of the exponential function and the monotone convergence theorem.

**Lemma 2.8** Let $y_1, y_2, \ldots$ be i.i.d. Assume that $\mu = Ey_1$ exists and for any real $b$ define

$$\begin{aligned} t &= \text{first } n \geq 1 \text{ such that } s_n \geq b \\ &= \infty \text{ if no such } n \text{ exists.} \end{aligned} \tag{2.46}$$

If $P(y_1 < b) > 0$ and $P(y_1 = 0) < 1$, then $P(t < \infty) = 1$ if and only if $\mu \geq 0$.

***Proof*** With no loss of generality we may assume that $b \geq 0$. If $0 < \mu \leq \infty$ the fact that $P(t < \infty) = 1$ follows at once from the strong law of large numbers (Theorem 1.6–see also Section 7.2(c)). If $\mu = 0$ the Chung-Fuchs Theorem (Section 2.4(e)) implies that $\limsup_{n\to\infty} s_n = +\infty$, and thus $P(t < \infty) = 1$ holds in this case also. Suppose next that $-\infty \leq \mu < 0$ and by way of contradiction that $P(t < \infty) = 1$. Then by defining $t_1, t_2, \ldots$ as in Section 2.2(e) we conclude that $P\{\limsup_n s_n = +\infty\} = 1$, contradicting the strong law of large numbers, which implies that $P\{\lim_n s_n = -\infty\} = 1$.

***Remark*** If $Ey_1^2 < \infty$ the case $\mu = 0$ is a straightforward application of the central limit theorem and the Kolmogorov $0-1$ law. This line of argument is developed in a different context in Section 4.7.

**Theorem 2.6** Let $y_1, y_2, \ldots$ be i.i.d. with $\varphi(\lambda) < \infty$ for some $\lambda \neq 0$. Let $D$ denote the interval of values $\lambda$ for which $\varphi(\lambda) < \infty$, and let $t$ be defined by (2.46). If $P(y_1 < b) > 0$, then for any $\lambda \in D$ (2.43) holds if and only if $\varphi'(\lambda) \geq 0$.

***Proof*** First observe that $Q\{y_1 < b\} > 0$ if and only if $P\{y_1 < b\} > 0$. By Lemma 2.8 $Q\{t < \infty\} = 1$ if and only if $\int_{-\infty}^{\infty} x\, dG(x) \geq 0$, which by Lemma 2.7(b) is equivalent to $\varphi'(\lambda) \geq 0$.

**Corollary (cf. problem 6)** Let $y_1, y_2, \ldots$ be i.i.d. with $\varphi(\lambda) < \infty$ for some $\lambda \neq 0$. For any real numbers $a, b$ ($a < b$) let $t =$ first $n \geq 1$ such that $s_n \leq a$ or $s_n \geq b$. Then (2.43) holds for every $\lambda$ for which $\varphi(\lambda) < \infty$.

***Proof*** Let

$$
\begin{aligned}
t_1 &= \text{first } n \geq 1 \text{ such that } s_n \geq b \\
&= \infty \text{ if no such } n \text{ exists} \\
t_2 &= \text{first } n \geq 1 \text{ such that } s_n \leq a \\
&= \infty \text{ if no such } n \text{ exists.}
\end{aligned}
$$

Then $t = \min(t_1, t_2)$. If $P\{y_1 \in (a, b)\} = 0$, the result is trivially true. Otherwise it follows immediately from the theorem, since

$$
Q\{t > n\} \leq \min\left(Q\{t_1 > n\}, Q\{t_2 > n\}\right), \tag{2.47}
$$

and for any $\lambda$ for which $\varphi(\lambda) < \infty$ at least one of the terms on the right-hand side of (2.47) tends to 0 as $n \to \infty$.

## 7. Application to the Sequential Probability Ratio Test

Let $f_0$ and $f_1$ be two probability density functions with respect to some $\sigma$-finite measure $\mu$, and for convenience assume that for any Borel set $A$, $\int_A f_0\, d\mu = 0$ if and only if $\int_A f_1\, d\mu = 0$. Suppose that $y_1, y_2, \ldots$ are i.i.d. with unknown density $f$. To test the hypothesis $H_0: f = f_0$ against the alternative $H_1: f = f_1$ A. Wald [1] proposed the following procedure: Let $f_{in} = \prod_{k=1}^{n} f_i(y_k)$ $(i = 0, 1; n = 1, 2, \ldots)$. Take positive numbers $A < 1 < B$ and define $t$ = first $n \geq 1$ such that $\frac{f_{1n}}{f_{0n}} \geq B$ or $\leq A$. Accept $H_0$ or $H_1$ according as $\frac{f_{1t}}{f_{0t}}$ is $\leq A$ or $\geq B$.

Let $P_i$ denote the probability measure under which the $y$'s have probability density function $f_i$ for $i = 0, 1$. By Section 2.2(d)

$$P_0\left\{\frac{f_{1m}}{f_{0m}} \to 0\right\} = 1 = P_1\left\{\frac{f_{1m}}{f_{0m}} \to \infty\right\}.$$

Hence $P_i\{t < \infty\} = 1$ $(i = 0, 1)$. Letting $\alpha$ and $\beta$ denote the error probabilities of the first and second kind respectively, we obtain from (2.42)

$$\begin{aligned}(2.48)\qquad \alpha \equiv P_0\{f_{1t} \geq Bf_{0t}\} &= \int_{\{f_{1t} \geq Bf_{0t}\}} \frac{f_{0t}}{f_{1t}}\, dP_1 \\ &\leq B^{-1}P_1\{f_{1t} \geq Bf_{0t}\} = B^{-1}(1 - \beta)\end{aligned}$$

and

$$\begin{aligned}(2.49)\qquad \beta \equiv P_1\{f_{1t} \leq Af_{0t}\} &= \int_{\{f_{1t} \leq Af_{0t}\}} \frac{f_{1t}}{f_{0t}}\, dP_0 \\ &\leq AP_0\{f_{1t} \leq Af_{0t}\} = A(1 - \alpha).\end{aligned}$$

"Neglecting the excess" of $f_{1t}/f_{0t}$ over the boundaries $A$ and $B$ and thus regarding the inequalities (2.48) and (2.49) as approximate equalities, we have

$$(2.50)\qquad \alpha \cong \frac{1 - A}{B - A}, \quad \beta \cong \frac{A(B - 1)}{B - A}.$$

Wald also suggested applying his sequential probability ratio test to problems involving certain composite hypotheses, and consequently it is of interest to investigate the behavior of $t$ and $f_{1t}/f_{0t}$ under probability measures $P$ such that $y_1, y_2, \ldots$ are i.i.d. random variables, but $P$ is not necessarily $P_0$ or $P_1$. By problem 6, $P\{t < \infty\} = 1$. If there exists a non-zero value $\lambda_1$ such that

$$E\left[\exp\left(\lambda_1 \log \frac{f_1(y_1)}{f_0(y_1)}\right)\right] = 1$$

(Lemma 2.6(d) gives a sufficient condition that such a value $\lambda_1$ exists), then putting $a = \log A$, $b = \log B$, $u_n = \log(f_1(y_n)/f_0(y_n))$, and $s_n = \sum_1^n u_k$ and as above neglecting the excess of $s_t$ over the boundaries $a$ and $b$, we have from the corollary to Theorem 2.4 that

$$1 = E(e^{\lambda_1 s_t}) \cong e^{\lambda_1 b} P\{s_t \geq b\} + e^{\lambda_1 a} P\{s_t \leq a\}.$$

Hence $P\{\text{reject } H_0\} = P\{f_{1t} \geq Bf_{0t}\} \cong \dfrac{1 - A^{\lambda_1}}{B^{\lambda_1} - A^{\lambda_1}}$. In particular for $P = P_0$, $\lambda_1 = 1$ and hence

$$P_0\{f_{1t} \geq Bf_{0t}\} \cong \frac{1 - A}{B - A},$$

while for $P = P_1$, $\lambda_1 = -1$ and

$$P_1\{f_{1t} \leq Af_{0t}\} \cong 1 - \frac{1 - A^{-1}}{B^{-1} - A^{-1}} = \frac{A(B - 1)}{B - A},$$

in agreement with (2.50).

A similar application of Theorem 2.3 to the martingale $s_n - nEu_1$ ($s_n^2 - nEu_1^2$ if $Eu_1 = 0$) yields approximations to $Et$, the expected number of observations required. (See problems 1 and 13.)

# PROBLEMS

**1.** *Gambler's ruin.* Let $y_1, y_2, \ldots$ be i.i.d., $P\{y_1 = 1\} = p$, $P\{y_1 = -1\} = q = 1 - p$, and let $s_n = \sum_1^n y_k$ $(n \geq 1)$. For any positive integers $a$, $b$ let

$$t = \text{first } n \quad \text{such that} \quad s_n = -a \text{ or } +b.$$

Show that

$$P\{s_t = -a\} = \begin{cases} \dfrac{1 - (p/q)^b}{1 - (p/q)^{a+b}} & (p \neq q) \\[2ex] \dfrac{b}{a + b} & (p = q = \frac{1}{2}) \end{cases}$$

and

$$Et = \begin{cases} \dfrac{b}{p - q} - \dfrac{a + b}{p - q} \cdot \dfrac{1 - (p/q)^b}{1 - (p/q)^{a+b}} & (p \neq q) \\[2ex] ab & (p = q = \frac{1}{2}). \end{cases}$$

(*Hint:* If $p \neq q$, $(q/p)^{s_n}$ and $s_n - n(p - q)$ form martingales relative to the appropriate $\sigma$-algebras, the former is a special case of $e^{\lambda s_n}/(\varphi(\lambda))^n$ at the point $\lambda = \lambda_1$. If $p = q$ the appropriate martingales are $s_n$ and $s_n^2 - n$.)

**2.** *Continuation.* Now let

$$t = \text{first } n \quad \text{such that} \quad s_n \geq b \\ = \infty \quad \text{if no such } n \text{ exists.} \tag{2.51}$$

(a) If $p < q$, $P\{t < \infty\} = (p/q)^b$.
(b) The generating function of $t$ is given by

$$\sum_1^\infty z^n P(t = n) = \left(\frac{1 - \sqrt{1 - 4pqz^2}}{2qz}\right)^b \quad (0 < z \leq 1). \tag{2.52}$$

(c) Show that if $p > q$ then $Et = \dfrac{b}{\mu}$, $\operatorname{Var} t = \dfrac{\sigma^2 b}{\mu^3}$, where $\mu = p - q$, $\sigma^2 = 1 - (p - q)^2$. (A direct proof is possible using the ideas of Theorems 2.4 and 2.5. Alternatively one can differentiate (2.52).)

(d) If $p > q$, then $\dfrac{t - (b/\mu)}{\sqrt{\sigma^2 b/\mu^3}}$ is approximately normally distributed as $b \to \infty$.

(e) Which of (a), (b), and (c) can be obtained by solving problem 1 and then letting $a \to -\infty$?

**3.** Let $y_1, y_2, \ldots$ be i.i.d. with distribution function $F$ for which

$$dF(x) = ce^{-\alpha x}\,dx \qquad (0 < x < \infty),$$

where $c > 0$, $\alpha > 0$. Then with $t$ defined by (2.51)

$$\sum_1^\infty z^n P(t = n) = \frac{\alpha - \lambda}{\alpha} e^{-\lambda b},$$

where for each $z$ $(0 < z \leq 1)$ $\lambda = \lambda(z)$ is the unique number in $[0, \alpha)$ satisfying $z\varphi(\lambda) = 1$.

**4.** Let $y_1, y_2, \ldots$ be i.i.d. with negative mean $\mu$. Assume that $\varphi(\lambda) < \infty$ for all $\lambda$ and let $t$ be defined by (2.51). Show that

$$P\{t < \infty\} \leq e^{-\lambda_1 b}, \tag{2.53}$$

where $\lambda_1$ satisfies $\varphi(\lambda_1) = 1$. When is there equality in (2.53)? Can you find $P\{t < \infty\}$ *exactly* for the distribution of problem 3?

**5.** Let $y_1, y_2, \ldots$ be i.i.d. with positive mean $\mu$ and with $\varphi(\lambda) < \infty$ for some $\lambda < 0$. Let $t$ be defined by (2.51). Show that for some $\lambda > 0$ $Ee^{\lambda t} < \infty$. (*Hint:* First consider the special case $y_i \leq a < \infty$.)

**6.** Give a direct proof of the corollary to Theorem 2.4. More generally, show from first principles that if $y_1, y_2, \ldots$ are i.i.d. with $P\{y_1 = 0\} < 1$, and if $N$ = first $n$ such that $s_n \geq b$ or $\leq a$, then $P\{N < \infty\} = 1$. (*Hint:* If $P\{|y_1| \geq b - a\} > 0$, then $N \leq N^* =$ first $n$ such that $|y_n| \geq b - a$. In general there exists an integer $r$ such that $P\{|s_r| \geq b - a\} > 0$.)

*Problems 7–11 suggest alternative proofs of the martingale convergence theorem. The cases of martingales indexed by $\{\ldots -1, 0\}$ and $\{1, 2, \ldots\}$ are treated by different methods. In problem 8 we assume that Lévy's Theorem (Theorem 1.4) is known.*

**7.** Let $\{x_n, \mathscr{F}_n, -\infty < n \leq 0\}$ be a martingale.

(a) For any $A \in \mathscr{F}_{-\infty} \equiv \bigcap_{-\infty}^{0} \mathscr{F}_n$ and $-\infty < s < \infty$

$$sP(A\{\max_{n \leq 0} x_n > s\}) \leq \int_{A\{\max_{n \leq 0} x_n > s\}} x_0.$$

(This is essentially Kolmogorov's inequality for martingales (2.16).)
(b) Applying (a) to $x_n$ and $-x_n$ conclude that if $r, s$ are rational $(r < s)$ and

$$A = \{\limsup_{n \to -\infty} x_n > s > r > \liminf_{n \to -\infty} x_n\},$$

then $P(A) = 0$. Hence $\lim_{n \to -\infty} x_n$ exists a.s.

**8.** Let $\{x_n, \mathscr{F}_n, 1 \leq n < \infty\}$ be a martingale with $\sup_n E|x_n| < \infty$.
(a) Let $z_n^{(1)} = \sup_{k \geq n} E(x_k^+ \mid \mathscr{F}_n)$, $z_n^{(2)} = z_n^{(1)} - x_n$. Then $\{z_n^{(i)}, \mathscr{F}_n, 1 \leq n < \infty\}$ is a non-negative martingale $(i = 1, 2)$, and $x_n = z_n^{(1)} - z_n^{(2)}$. Hence to show that $\lim_{n \to \infty} x_n$ exists we may without loss of generality assume that $x_n \geq 0$ $(n = 1, 2, \ldots)$. By passing to $\min(x_n, b)$ we may in fact assume that $\{x_n, \mathscr{F}_n, 1 \leq n < \infty\}$ is a uniformly bounded, non-negative supermartingale.
(b) Now assume that $\{x_n, \mathscr{F}_n, 1 \leq n < \infty\}$ is a uniformly bounded, non-negative supermartingale, and let $x_\infty = \limsup_{n \to \infty} x_n$. Then for every stopping time $t \geq n$

$$x_n \geq E(x_t \mid \mathscr{F}_n).$$

Apply to

$$t = \text{first } k \geq n \quad \text{such that} \quad x_k \geq E(x_\infty \mid \mathscr{F}_k) - \varepsilon$$

(by Lévy's Theorem $P\{t < \infty\} = 1$) to conclude

$$x_n \geq E(x_\infty \mid \mathscr{F}_n) \qquad (n = 1, 2, \ldots).$$

Apply Lévy's Theorem again to obtain

$$\liminf_{n\to\infty} x_n \geq x_\infty = \limsup_{n\to\infty} x_n.$$

**9.** Extend the results of problems 7 and 8 to submartingales by means of the representation (2.1).

**10.** Let $\{x_n, \mathscr{F}_n, 1 \leq n < \infty\}$ be a non-negative supermartingale. Show that for any $n = 1, 2, \ldots$ and $r < s$,

$$P\{\max_{k\geq n} x_k \geq s \mid \mathscr{F}_n\} \leq r/s \quad \text{on} \quad \{x_n \leq r\}.$$

**11.** *Application.* Let $\{x_n, \mathscr{F}_n, 1 \leq n < \infty\}$ be a non-negative supermartingale and for any $0 \leq r < s$ define $t_1, t_2, \ldots$ as in the proof of the upcrossing inequality. Show that

$$P\{t_{2m} < \infty\} \leq r/sP\{t_{2m-1} < \infty\}.$$

Conclude that $\lim_{n\to\infty} x_n$ exists a.s.

**12.** Give examples to show that neither of the conditions appearing in (2.13) implies the other.

**13.** In the notation of Section 2.7 find an approximation to $P\{f_{1t} \geq Bf_{0t}\}$ which is valid when $Eu_1 = 0$. Find an approximation to $Et$, the expected number of observations required. (*Hint:* Look again at problem 1.)

**14.** Let $y_1, y_2, \ldots$ be independent r.v.'s with $Ey_k = 0$, $Ey_k^2 = \sigma_k^2$ $(k = 1, 2, \ldots)$. Use the Hájek-Rényi inequality (Section 2.4(b)) to show that if $\sum_1^\infty \sigma_k^2/k^2 < \infty$, then $n^{-1}\left(\sum_1^n y_k\right) \to 0$.

# Chapter 3

# Introductory Theory

We are now in a position to embark on the study of the optimal stopping problem. We begin by giving a formal statement of the problem. Several examples, which will reappear from time to time, are presented next. We then discuss the *finite* and *monotone* cases, which can be more or less completely solved without recourse to the general theory of Chapter 4. The general theory does, of course, contribute to one's understanding of these special problems; and conversely, the results presented in this chapter motivate to some extent our subsequent approach to the general problem.

## 1. Statement of the Problem and Examples

Assume that we can observe sequentially random variables $y_1, y_2, \ldots,$ having a known joint distribution. We must stop the observation process at some point, and if we stop at the $n$th stage, we receive a "reward" $x_n$, a known function of $y_1, \ldots, y_n$. We are interested in finding stopping rules which maximize our expected reward.

More formally, we assume that there is given (i) a probability space $(\Omega, \mathscr{F}, P)$, (ii) an increasing sequence $(\mathscr{F}_n)$ of sub-$\sigma$-algebras of $\mathscr{F}$, and (iii) a sequence $x_1, x_2, \ldots$ of random variables such that $x_n$ is measurable with respect to $\mathscr{F}_n$, $n = 1, 2, \ldots.$ Such a pair of sequences $\{x_n, \mathscr{F}_n\}_1^\infty$ is called a *stochastic sequence*. A *stopping rule* or *stopping variable* (s.v.) is a random variable $t$ with values $1, 2, \ldots, +\infty$, such that $P(t < \infty) = 1$ and $\{t = n\} \in \mathscr{F}_n$, $n = 1, 2, \ldots.$ (In terms of the intuitive background of the preceding paragraph $\mathscr{F}_n = \mathscr{B}(y_1, \ldots, y_n)$, and it is convenient to keep this interpretation in mind although our general results do not depend on it. With this interpretation the requirement that $\{t = n\} \in \mathscr{F}_n$ means that our decision to stop at time $n$ must be made on the basis of the already observed $y_1, \ldots, y_n$ and must not involve future values $y_{n+1}, \ldots$.) If $\{x_n, \mathscr{F}_n\}_1^\infty$ is any stochastic sequence and $t$ any s.v., then the random variable $x_t$ defined by

$$x_t = \sum_{n=1}^{\infty} x_n I_{\{t=n\}} = \begin{cases} x_n & \text{on } \{t = n\} \quad \text{for } n = 1, 2, \ldots \\ 0 & \text{on } \{t = \infty\} \end{cases}$$

is our actual reward, and we define the *value* $V$ of the stochastic sequence $\{x_n, \mathscr{F}_n\}_1^\infty$ to be $\sup Ex_t$, where the supremum is taken over all s.v.'s $t$ such that $Ex_t$ exists. Since for an arbitrary stochastic sequence it is not always clear whether there exist rules $t$ such that $Ex_t$ exists, we shall find it convenient to assume that $\{x_n, \mathscr{F}_n\}_1^\infty$ is an *integrable* stochastic sequence; i.e., that $E|x_n| < \infty$ $(n \geq 1)$. Then we see that $-\infty < Ex_1 \leq V \leq \infty$. We shall be interested in answering the following questions:

(a) How can we compute $V$?
(b) Is there an *optimal* s.v. $t$, i.e., one for which $Ex_t$ exists and equals $V$?
(c) If an optimal rule exists, what is its nature?

On occasion it will be convenient to discuss the optimal stopping problem relative to some subclass of the class of all s.v.'s. Thus if $D$ is any class of s.v.'s $t$ such that $Ex_t$ exists, we define $V(D) = \sup_{t \in D} Ex_t$, and say that $t$ is optimal in $D$ if $t \in D$ and $Ex_t = V(D)$. Let $C$ denote the class of all s.v.'s $t$ such that $Ex_t^- < \infty$. It is clear that $V = V(C)$.

## Examples

(a) $\Omega$ has only one element; $x_n = 1 - 1/n$, $n = 1, 2, \ldots$. Clearly $V = 1$ but no optimal rule exists.
(b) Let $(y_n)$ be i.i.d., $P(y_n = 1) = p = 1 - P(y_n = 0)$, and for $n = 1, 2, \ldots$ let $\mathscr{F}_n = \mathscr{B}(y_1, \ldots, y_n)$, $x_n = (y_1 + \cdots + y_n)/n$. It is easy to see that $V \geq p(2 - p)$. For, given any $\varepsilon > 0$ let

$$\begin{aligned} t &= 1 \quad \text{if } y_1 = 1 \\ &= \inf\{n: x_n \geq p - \varepsilon\} \quad \text{if } y_1 = 0. \end{aligned}$$

The strong law of large numbers shows that $t$ is a s.v., and $V \geq Ex_t \geq p + (1 - p)(p - \varepsilon)$. Letting $\varepsilon \to 0$, the assertion follows. It is not obvious, however, what the exact value of $V$ is (in fact, no formula for $V$ as a function of $p$ is known), that an optimal rule exists (it does), or even that $V$ is an increasing function of $p$ (it is).

(c) Let $\{x_n, \mathscr{F}_n, 1 \le n < \infty\}$ be a supermartingale (for definition see Section 2.1) for which $Ex_1 < \infty$. Then by Theorem 2.2

$$Ex_t \le Ex_1 \tag{3.1}$$

for every s.v. $t$ such that for some $1 \le N < \infty, P(t \le N) = 1$. Hence if $D$ denotes the class of bounded s.v.'s,

$$V(D) = Ex_1 \quad \text{and} \quad t = 1 \text{ is optimal in } D.$$

If $(x_n^-)$ is uniformly integrable, Theorem 2.2 shows that (3.1) holds for every s.v. $t$.

(d) *Continuation.* Let $y_1, y_2, \ldots$ be i.i.d. with mean $\mu$, and let $\mathscr{F}_n = \mathscr{B}(y_1, \ldots, y_n)$, $x_n = y_1 + \cdots + y_n$. If $\mu \le 0$, then $\{x_n, \mathscr{F}_n, 1 \le n < \infty\}$ is a supermartingale and the preceding remarks apply. If

$$P\{y_n = 1\} = \tfrac{1}{2} = P\{y_n = -1\}, \quad \text{then } \mu = 0, \quad \text{but } V = +\infty,$$

so (3.1) is not generally true. However, we can give a complete solution to the optimal stopping problem for $\{x_n, \mathscr{F}_n\}_1^\infty$ by a relatively simple direct argument.

**Theorem 3.1** Let $(y_n)$ be i.i.d. with expectation $\mu$ and for $n = 1, 2, \ldots$ let $\mathscr{F}_n = \mathscr{B}(y_1, \ldots, y_n)$, $x_n = y_1 + \cdots + y_n$. Then
(i) if $\mu < 0$, $V = Ex_1$ and $t = 1$ is optimal;
(ii) if $\mu \ge 0$, then $V = \infty$ except when the $y$'s are identically 0, in which case $V = 0$.

***Proof*** Part (ii) follows from Lemma 2.8. Part (i) follows at once from the following generalization of Wald's lemma. (Compare with Theorem 2.3.)

**Lemma 3.1** Let $(y_n)$ be i.i.d. with expectation $\mu$, and for $n = 1, 2, \ldots$ let $\mathscr{F}_n = \mathscr{B}(y_1, \ldots, y_n)$, $x_n = y_1 + \cdots + y_n$. Then for any s.v. $t$ such that $Ex_t$ exists, $Ex_t = \mu Et$ provided it is not the case that both $\mu = 0$ and $Et = \infty$. In order that $Ex_t$ exist and $= \mu Et$ it suffices that $\mu$ and $Et$ both be finite.

***Proof*** Without loss of generality we may assume that $\Omega$ is the space of sequences $\omega = (y_1, y_2, \ldots)$ and $y_n(\omega) = y_n$ $(n \ge 1)$. Define a transformation $Q$ on $\Omega$ by

$$Q(\omega) = (y_{t+1}(\omega), y_{t+2}(\omega), \ldots).$$

Let $t_0 = 0$, $t_1(\omega) = t(\omega)$, $t_2(\omega) = tQ(\omega), \ldots, t_n(\omega) = tQ^{n-1}(\omega)$, $\ldots$, where $Q^1 = Q$, $Q^{n+1} = Q(Q^n)$. Defining $z_i$ to be

$$y_{t_1+\cdots+t_{i-1}+1} + \cdots + y_{t_1+\cdots+t_i}$$

for $i = 1, 2, \ldots$, we easily see that $(t_1, z_1)$, $(t_2, z_2), \ldots$ are i.i.d. vectors. Assume that $Ex_t = Ez_1$ exists. Since

$$\text{(3.2)} \quad n^{-1}\left(\sum_1^n z_i\right) = \left(\frac{y_1 + \cdots + y_{t_1+\cdots+t_n}}{t_1 + \cdots + t_n}\right)\left(\frac{t_1 + \cdots + t_n}{n}\right),$$

the first part of the lemma follows from the strong law of large numbers. If $\mu$ and $Et$ are both finite, then by the above argument the right-hand side of (3.2) converges to a finite number, and the strong law now asserts that $Ez_1 = Ex_t$ exists and is finite.

(e) Imagine a burglar who each day loots some house. His daily gains form a sequence of i.i.d. random variables with finite expectation. Each day there is a positive probability $p$ of his being caught and forced to return his entire profit. We regard the event that the burglar is apprehended on the $n$th day as independent of all that has occurred in the past. The question is that of choosing most wisely the time for the burglar to retire.

More formally, let $y_1, y_2, \ldots$ be a sequence of i.i.d. non-negative r.r.v.'s with finite expectation. Let $\delta_1, \delta_2, \ldots$ be independent random variables such that $p = P\{\delta_n = 1\} = 1 - P\{\delta_n = 0\}$ for $n = 1, 2, \ldots$, and assume that the $y$'s and the $\delta$'s are independent. Let $\mathscr{F}_n = \mathscr{B}(y_1, \ldots, y_n, \delta_1, \ldots, \delta_n)$. The burglar would like to solve the optimal stopping problem for $\{x_n, \mathscr{F}_n\}_1^\infty$, where

$$x_n = (\delta_1 \delta_2 \cdots \delta_n)(y_1 + \cdots + y_n).$$

(f) An executive is faced with the problem of hiring a secretary from a group of $N$ girls. He can interview the girls one at a time, after each interview accepting or rejecting that particular girl. If he rejects a girl, he may not recall her. At any stage he knows how the girl he is now interviewing ranks compared with her predecessors, but not how she compares with the girls he has not yet seen. We assume that he sees the girls in a random order and ask how he should go about deciding which to hire. In particular we might ask for a rule such that the probability of choosing the best of the $N$ girls is a maximum or, on the other hand, such that the expected absolute rank of the girl chosen is a minimum (1 being the rank of the best girl, 2 the rank of the next best, $\ldots$, $N$ the rank of the worst).

Formally, let $a_1, a_2, \ldots, a_N$ denote a permutation of the integers $1, 2, \ldots, N$, all permutations being equally likely. The integer 1 corresponds to the best girl, ..., $N$ to the worst. For any $n = 1, \ldots, N$ let $y_n$ = number of terms $a_1, \ldots, a_n$ which are $\leq a_n$ ($y_n$ = relative rank of the $n$th girl to appear), and let $\mathscr{F}_n = \mathscr{B}(y_1, \ldots, y_n)$. If the criterion is to be that of maximizing the probability of choosing the best girl, the reward is 1 if the $n$th girl is selected and $a_n = 1$ $(n = 1, \ldots, N)$, and 0 otherwise. A minor technical difficulty to be overcome here is that such a reward sequence does not satisfy the measurability requirement of a stochastic sequence, i.e., the random variable which is 1 if $a_n = 1$ and 0 otherwise is *not* $\mathscr{F}_n$-measurable for $n < N$. But, putting $x_n = P(a_n = 1 \mid \mathscr{F}_n)$, $n = 1, \ldots, N$, $x_n$ is $\mathscr{F}_n$-measurable, and for any s.v. $t$,

$$Ex_t = \sum_1^N \int_{\{t=n\}} x_n = \sum_1^N \int_{\{t=n\}} P(a_n = 1 \mid \mathscr{F}_n)$$
$$= \sum_1^N \int_{\{t=n\}} I\{a_n = 1\} = P\{a_t = 1\}.$$

Hence choosing a strategy to maximize the probability of getting the best girl is equivalent to solving the optimal stopping problem for $\{x_n, \mathscr{F}_n\}_1^N$. On the other hand, if our criterion is to be that of minimizing the expected absolute rank of the girl selected we would put

$$\tilde{x}_n = -E(a_n \mid \mathscr{F}_n)$$

and solve the optimal stopping problem for $\{\tilde{x}_n, \mathscr{F}_n\}_1^N$.

(g) We are trying to park an automobile close to a desired location designated by the point 0. We approach our goal along the negative half line and for $n = \ldots, -2, -1, 0, 1, \ldots$, the $n$th parking spot is unoccupied with probability $p$ independently of the other places. If a spot is occupied we are not permitted to park there, whereas if we park in an unoccupied spot we incur a "loss" proportional to the distance of this spot from our goal, 0.

For simplicity we assume that there is a lower bound $Q < 0$ on the permissible parking spots. Let $y_Q, y_{Q+1}, \ldots$ be i.i.d. such that $P(y_n = 1) = p = 1 - P(y_n = 0)$, and for $n = Q, Q + 1, \ldots$ let

$$\mathscr{F}_n = \mathscr{B}(y_Q, y_{Q+1}, \ldots, y_n),$$

$$\tilde{x}_n = \begin{cases} -\infty & \text{if} \quad y_n = 0 \\ -|n| & \text{if} \quad y_n = 1. \end{cases}$$

Since $\tilde{x}_n \leq 0$ for all $n$ there is no question about the existence of $E\tilde{x}_t$ for any s.v. $t$. However, the stochastic sequence $\{\tilde{x}_n, \mathscr{F}_n\}_Q^\infty$ does not fulfill our formal requirement that $E|\tilde{x}_n| < \infty$ $(n \geq Q)$. Disregarding this difficulty momentarily, we see that in trying to maximize $E\tilde{x}_t$ we may restrict our consideration to rules with the property that

$$t = t_1 \quad \text{if} \quad t > 0,$$

where $t_1$ = first $n \geq 1$ such that $y_n = 1$. In fact for any s.v. $t$,

$$t' = tI(t \leq 0) + t_1 I(t > 0)$$

is a s.v. with the property that $\tilde{x}_{t'} \geq \tilde{x}_t$. It is easily seen that $E\tilde{x}_{t_1} = -1/p$. We do not change the original problem then if we put

$$x_1 = -1/p,$$

$$x_n = \begin{cases} n & \text{if} \quad y_n = 1 \\ -1/p & \text{if} \quad y_n = 0, \end{cases} \qquad Q \leq n \leq 0;$$

and the integrable stochastic sequence $\{x_n, \mathscr{F}_n\}_Q^1$ defines an optimal stopping problem fitting into our general scheme.

(h) The following example has motivated a considerable amount of research in theoretical statistics. We shall present a heuristic solution to begin with and a rigorous one later in Sections 4.3(a) and 5.2(a).

Let $(y_n)$ be i.i.d. random variables with density function $f$ with respect to some $\sigma$-finite measure $\mu$ on the Borel sets of the line. It is desired to test the (simple) hypothesis $H_0: f = f_0$ versus $H_1: f = f_1$ where $f_0$ and $f_1$ are two specified densities. The loss due to accepting $H_0$ when $H_1$ is true is some constant $b > 0$ and that due to accepting $H_1$ when $H_0$ is true is $a > 0$; the cost of making each observation $y_n$ is unity. A sequential decision procedure $(\delta, t)$ provides for determining a sample size $t$ and making a terminal decision $\delta$; the expected loss for $(\delta, t)$ is

$$a\alpha_0 + E_0 t \quad \text{when } H_0 \text{ is true,}$$

$$b\alpha_1 + E_1 t \quad \text{when } H_1 \text{ is true,}$$

where $\alpha_0 = P_0(\text{accept } H_1)$, $\alpha_1 = P_1(\text{accept } H_0)$.

If there is an a priori probability $\pi$ that $H_0$ is true (and hence probability $1 - \pi$ that $H_1$ is true) the global "risk" for $(\delta, t)$ is given by

$$r(\pi, \delta, t) = \pi[a\alpha_0 + E_0 t] + (1 - \pi)[b\alpha_1 + E_1 t].$$

For a given s.v. $t$ it is easy to determine a terminal decision rule $\delta$ which minimizes $r(\pi, \delta, t)$ for fixed values of $a$, $b$, and $\pi$. For, the part of $r(\pi, \delta, t)$ which depends on $\delta$ is (omitting symbols like $d\mu(y_1) \cdots d\mu(y_n)$)

$$\begin{aligned}
\pi a \alpha_0 &+ (1 - \pi) b \alpha_1 \\
&= \pi a \sum_1^\infty \int_{\{t=n,\ \text{accept } H_1\}} f_0(y_1) \cdots f_0(y_n) \\
&\quad + (1 - \pi) b \sum_1^\infty \int_{\{t=n,\ \text{accept } H_0\}} f_1(y_1) \cdots f_1(y_n) \\
&\geq \sum_1^\infty \int_{\{t=n\}} \min\,[\pi a f_0(y_1) \cdots f_0(y_n),\ (1 - \pi) b f_1(y_1) \cdots f_1(y_n)] \\
&= \sum_1^\infty \int_{\{t=n\}} \min\,[a\pi_n,\ b(1 - \pi_n)][\pi f_0(y_1) \cdots f_0(y_n) \\
&\qquad\qquad + (1 - \pi) f_1(y_1) \cdots f_1(y_n)],
\end{aligned}$$

where

$$\begin{aligned}
\pi_n &= \pi_n(y_1, \ldots, y_n) \\
&= \frac{\pi f_0(y_1) \cdots f_0(y_n)}{\pi f_0(y_1) \cdots f_0(y_n) + (1 - \pi) f_1(y_1) \cdots f_1(y_n)}.
\end{aligned}$$

For a given s.v. $t$ define $\delta'$ by

$$\begin{cases} \text{accept } H_1 & \text{if} \quad t = n \quad \text{and} \quad \pi_n a \leq (1 - \pi_n) b, \\ \text{accept } H_0 & \text{if} \quad t = n \quad \text{and} \quad \pi_n a > (1 - \pi_n) b. \end{cases}$$

Then

$$\pi a \alpha_0(\delta, t) + (1 - \pi) b \alpha_1(\delta, t) \geq \pi a \alpha_0(\delta', t) + (1 - \pi) b \alpha_1(\delta', t).$$

Hence finding a pair $(\delta, t)$ which for given $\pi$ minimizes $r(\pi, \delta, t)$ (the "Bayes" decision procedure) amounts to solving the following problem: for given $0 < \pi < 1$ let $y_1, y_2, \ldots$ have the joint density function for each $n$ equal to

$$\pi f_0(y_1) \cdots f_0(y_n) + (1 - \pi) f_1(y_1) \cdots f_1(y_n),$$

where $f_0$ and $f_1$ are given univariate density functions. For given $a$, $b > 0$, let

$$h(\lambda) = \min\,(a\lambda, b(1 - \lambda)), \qquad 0 \leq \lambda \leq 1,$$

$$\begin{cases} \pi_0 = \pi \\ \pi_n = \dfrac{\pi f_0(y_1)\cdots f_0(y_n)}{\pi f_0(y_1)\cdots f_0(y_n) + (1-\pi)f_1(y_1)\cdots f_1(y_n)}, \\ \qquad\qquad\qquad\qquad\qquad\qquad\qquad n = 1, 2, \ldots \\ x_n = -h(\pi_n) - n, \qquad\qquad n = 0, 1, \ldots \\ \mathscr{F}_0 = \{\phi, \Omega\} \\ \mathscr{F}_n = \mathscr{B}(y_1, \ldots, y_n), \qquad\qquad n = 1, 2, \ldots. \end{cases}$$

We want to find a s.v. $t$ such that $Ex_t$ is a maximum. (Note that in the present context we are allowed to take no observations on the $y_n$ and to decide in favor of $H_0$ or $H_1$ with $x_0 = h(\pi)$; a s.v. which assumes the value 0 must do so with probability 0 or 1, since $\mathscr{F}_0 = \{\phi, \Omega\}$. This comment is easily incorporated into the above observations concerning optimal terminal decision rules.) The problem is trivial if $a \leq 1$ or $b \leq 1$ since then $h(\lambda) < 1$ and $x_0 > x_n$ for all positive $n$, so that the optimal rule is $t = 0$. We thus assume that $a > 1, b > 1$.

The following non-rigorous argument has considerable intuitive appeal. We ask initially whether we should take a first observation. We compute

$$(3.3) \quad V(\pi) = \inf_{\delta,t} [\pi(\alpha_0 a + E_0 t) + (1-\pi)(\alpha_1 b + E_1 t)],$$

where the inf is taken over all s.v.'s $t$ which take at least one observation and over all terminal decision rules $\delta$. $V(\pi)$ then represents our minimum expected loss if we take at least one observation. If we take no observations our loss is $h(\pi)$; and hence we should take a first observation if and only if $h(\pi) > V(\pi)$. We now make two comments about the sequence of rewards $(x_n)$: (i) $x_n$ depends on $y_1, \ldots, y_n$ only through the value of $\pi_n$, and (ii) writing

$$\pi_n = \frac{\pi_{n-1} f_0(y_n)}{\pi_{n-1} f_0(y_n) + (1-\pi_{n-1}) f_1(y_n)}, \qquad n = 1, 2, \ldots,$$

we see that the $\pi_n$ form a stationary Markov sequence in the sense that for $n = 0, 1, \ldots$ the conditional distribution of $\pi_{n+1}$ given $\mathscr{F}_n$ is a function of $\pi_n$ not depending on $n$. Suppose then that we take an initial observation. If we stop, our loss is $h(\pi_1) + 1$. If we continue, our prospects for the future are precisely the same as at stage 0 except that (a) we have already paid a dollar for the privilege of making a first observation, and (b) our a priori probability is now $\pi_1$.

Hence the minimum expected loss among rules taking at least one more observation is $V(\pi_1) + 1$, and we should take a second observation if and only if $h(\pi_1) + 1 > V(\pi_1) + 1$. The argument is repeated by induction, so that the natural candidate for an optimal s.v. is $\sigma = \inf \{n: h(\pi_n) \leq V(\pi_n)\}$. Let $A = \{\pi: h(\pi) \leq V(\pi)\}$. Since $V(\cdot)$ is the inf of a family of linear functions, it is concave; $V(0) = V(1) = 1$. Assuming that $A \neq [0, 1]$, it is easily seen that there are unique numbers $\pi'$ and $\pi''$ with $\pi' < \pi''$ such that $A = \{\pi: \pi \leq \pi' \text{ or } \pi \geq \pi''\}$, and that $(\delta', \sigma)$ is a Wald sequential probability ratio test (see Wald and Wolfowitz [1] or Lehmann [1], pp. 104–106).

## 2. The Finite Case. Backward Induction

In some cases, e.g. the secretary problem of Section 3.1 (f), there are only a finite number $N$ of random variables to be observed. Even in the infinite case it is of interest to see what can be achieved by stopping rules which are bounded by a fixed number $N$. In all such cases a complete solution is provided in principle by the method of backward induction to be described below.

Let $\{x_n, \mathscr{F}_n\}_1^\infty$ be an integrable stochastic sequence. We shall denote by $C^N$ the class of all s.v.'s $t$ such that $t \leq N$, and define

$$V^N = \sup_{t \in C^N} Ex_t.$$

(Note that $E|x_t| \leq \sum_{n=1}^{N} E|x_n| < \infty$ and hence $Ex_t$ exists for all $t \in C^N$.) In keeping with our previous terminology we shall say that $s^N$ is optimal in $C^N$ if $s^N \in C^N$ and $Ex_{s^N} = V^N$. It is convenient to imbed the problem of computing $V^N$ and $s^N$ in a family of problems as follows. For each $n = 1, 2, \ldots, N$, let $C_n^N$ denote the set of all s.v.'s $t$ such that $n \leq t \leq N$ and define $v_n^N = \sup_{t \in C_n{}^N} Ex_t$. (Note that $C^N = C_1^N$ and $V^N = v_1^N$.) We shall now show how to find a s.v. which is optimal in $C_n^N$. If $n = N$, the problem is trivial since $t = N$ is the only stopping rule in $C_N^N$, and hence $v_N^N = Ex_N$. For $n = N - 1$ it is intuitively clear that we should compare $x_{N-1}$ with $E(x_N \mid \mathscr{F}_{N-1})$ and use the rule

$$t = \begin{cases} N-1 & \text{if} \quad x_{N-1} \geq E(x_N \mid \mathscr{F}_{N-1}), \\ N & \text{if} \quad x_{N-1} < E(x_N \mid \mathscr{F}_{N-1}). \end{cases}$$

These considerations motivate the following "dynamic programming" theorem, which formalizes the principle of *backward induction*.

**Theorem 3.2** Let $N$ be a fixed positive integer. Define successively $\gamma_N^N, \gamma_{N-1}^N, \ldots, \gamma_1^N$ by setting

$$\gamma_N^N = x_N,$$
$$\gamma_n^N = \max\,[x_n, E(\gamma_{n+1}^N \mid \mathscr{F}_n)], \qquad n = N-1, \ldots, 1.$$

For each $n = 1, 2, \ldots, N$, let

$$s_n^N = \text{first } i \geq n \quad \text{such that} \quad x_i = \gamma_i^N.$$

Then $s_n^N \in C_n^N$ and

$$E(x_{s_n^N} \mid \mathscr{F}_n) = \gamma_n^N \geq E(x_t \mid \mathscr{F}_n),\ t \in C_n^N;$$

so that

$$Ex_{s_n^N} = E\gamma_n^N \geq Ex_t,\ t \in C_n^N, \quad \text{and} \quad v_n^N = E\gamma_n^N.$$

***Proof*** The theorem is trivial for $n = N$. Assume that it is true for some value $n = 2, 3, \ldots, N$. Take any $t \in C_{n-1}^N$ and any $A \in \mathscr{F}_{n-1}$; set $t' = \max(t, n) \in C_n^N$. Then

$$\int_A x_t = \int_{A(t=n-1)} x_{n-1} + \int_{A(t \geq n)} x_{t'}$$

$$(3.4) \qquad = \int_{A(t=n-1)} x_{n-1} + \int_{A(t \geq n)} E[E(x_{t'} \mid \mathscr{F}_n) \mid \mathscr{F}_{n-1}]$$

$$\leq \int_{A(t=n-1)} x_{n-1} + \int_{A(t \geq n)} E(\gamma_n^N \mid \mathscr{F}_{n-1}) \leq \int_A \gamma_{n-1}^N,$$

so $E(x_t \mid \mathscr{F}_{n-1}) \leq \gamma_{n-1}^N$. For $t = s_{n-1}^N$ we have $t' = s_n^N$ on $\{s_{n-1}^N \geq n\}$, and hence we may replace the inequalities in (3.4) by equalities. In fact, by the induction assumption $E(x_{s_n^N} \mid \mathscr{F}_n) = \gamma_n^N$, and thus by definition of $s_{n-1}^N$

$$\int_A x_{s_{n-1}^N} = \int_{A\{x_{n-1} \geq E(\gamma_n^N \mid \mathscr{F}_{n-1})\}} x_{n-1}$$

$$+ \int_{A\{x_{n-1} < E(\gamma_n^N \mid \mathscr{F}_{n-1})\}} E(\gamma_n^N \mid \mathscr{F}_{n-1})$$

$$= \int_A \max\,(x_{n-1}, E(\gamma_n^N \mid \mathscr{F}_{n-1}))$$

$$= \int_A \gamma_{n-1}^N, \quad \text{so} \quad E(x_{s_{n-1}^N} \mid \mathscr{F}_{n-1}) = \gamma_{n-1}^N.$$

## 3. An Application

The $\gamma_n^N$ of Theorem 3.2 can rarely be computed analytically by the backward induction which defines them. Nevertheless, there are occasional problems in which Theorem 3.2 allows a more or less explicit determination of the optimal stopping rule $s^N$.

Suppose that in the problem of Example 3.1 (f) we are interested in maximizing the probability of choosing the best girl. It is easy to see that $y_1, \ldots, y_N$ are independent and that

$$P\{y_n = j\} = 1/n, \qquad j = 1, 2, \ldots, n. \tag{3.5}$$

(3.6)

$$x_n = P\{a_n = 1 \mid \mathscr{F}_n\} = P\{a_n = 1 \mid y_n\} = \begin{cases} n/N & \text{if} \quad y_n = 1 \\ 0 & \text{if} \quad y_n > 1. \end{cases}$$

Using (3.5) and (3.6) and the method of backward induction introduced above, we have, dropping the fixed superscript $N$ and putting $v_n = E\gamma_n$,

$$\begin{aligned} \gamma_N &= x_N \\ \gamma_{N-1} &= \max\,(x_{N-1}, E(x_N)) = \max\,(x_{N-1}, v_N) \\ &\vdots \\ \gamma_n &= \max\,(x_n, E\gamma_{n+1}) = \max\,(x_n, v_{n+1}) \\ &\vdots \end{aligned}$$

and $s$ = first $n \geq 1$ such that $x_n \geq v_{n+1}$, where for the sake of consistency we put $v_{N+1} = 0$. Since $v_1 \geq v_2 \geq \cdots \geq v_N = Ex_N = 1/N > 0$, it follows from (3.6) that $s$ can be described as follows: there exists a number $r$, $1 \leq r \leq N$, such that $s$ = first $n \geq r$ such that $y_n = 1$. Straightforward calculations show that (aside from the trivial case $N = 1$)

$$Ex_s = [(r-1)/N] \sum_{k=r}^{N} (k-1)^{-1}. \tag{3.7}$$

Having applied Theorem 3.2 to restrict the class of rules within which we must look for an optimal rule, we find it convenient at this point to disregard the possibility of actually computing $s$ via backward induction and instead to maximize directly the right-hand side of (3.7) regarded as a function of $r$, say $\varphi(r)$. It is easily seen that

$$\varphi(r) - \varphi(r+1) = N^{-1}\left(1 - \sum_{r}^{N-1} k^{-1}\right), \tag{3.8}$$

and hence that $\varphi(r)$ is a maximum at

$$(3.9)\qquad r^* = r^*(N) = \inf\{r: \varphi(r) - \varphi(r+1) \geq 0\}$$
$$= \inf\left\{r: \frac{1}{r} + \frac{1}{r+1} + \cdots + \frac{1}{N-1} \leq 1\right\}.$$

We proceed to show that

$$(3.10)\qquad \lim_{N\to\infty} N^{-1}r^*(N) = \lim_{N\to\infty} v^N = e^{-1}.$$

In fact, it follows from (3.8) and (3.9) that for sufficiently large $N$

$$\sum_{r^*}^{N-1} k^{-1} \leq 1 < \sum_{r^*-1}^{N-1} k^{-1}$$

and hence

$$(3.11)\qquad \int_{r^*}^{N-1} y^{-1}\,dy \leq 1 < \int_{r^*-2}^{N-1} y^{-1}\,dy.$$

The fact that $\lim_N N^{-1}r^*(N)$ exists and equals $e^{-1}$ is a direct consequence of (3.11); the remainder of (3.10) now follows by an appropriate passage to the limit on the right-hand side of (3.7).

## 4. Some Fundamental Lemmas

Before beginning our discussion of the simplest infinite case, the so-called monotone case, we shall establish some basic general results. Lemma 3.3 is an extension of Theorem 2.2 appropriate to the optimal stopping problem. It plays a central role in all that follows. In fact Theorem 3.2 follows easily from it (see problem 8).

We are given an integrable stochastic sequence $\{x_n, \mathscr{F}_n\}_1^\infty$. Let $C$ denote the class of all s.v.'s $t$ for which $Ex_t^- < \infty$; then by definition $V = \sup_C Ex_t$.

**Lemma 3.2** If $s, t \in C$ and for each $n \geq 1$

$$(3.12)\qquad E(x_s \mid \mathscr{F}_n) \geq x_n \quad \text{on} \quad \{s > n\}$$

and

$$(3.13)\qquad E(x_t \mid \mathscr{F}_n) \leq x_n \quad \text{on} \quad \{s = n, t \geq n\},$$

then $Ex_s \geq Ex_t$. As a partial converse, if $Ex_s = V < \infty$, then for every $t \in C$, (3.12) and (3.13) are satisfied, $n = 1, 2, \ldots$.

***Proof***

$$Ex_s = \int_{(s<t)} x_s + \int_{(s\geq t)} x_s = \sum_1^\infty \int_{(s=n<t)} x_n + \sum_1^\infty \int_{(t=n\leq s)} E(x_s \mid \mathscr{F}_n)$$
$$\geq \sum_1^\infty \int_{(s=n<t)} E(x_t \mid \mathscr{F}_n) + \sum_1^\infty \int_{(t=n\leq s)} x_n = Ex_t.$$

Suppose now that $Ex_s = V < \infty$. To prove (3.12), let $n$ be arbitrary and

$$A = \{s > n, E(x_s \mid \mathscr{F}_n) < x_n\} \in \mathscr{F}_n.$$

Then $t' = nI_A + sI_{\bar{A}} \in C$, and if $P(A) > 0$,

$$Ex_{t'} = \int_A x_n + \int_{\bar{A}} x_s > \int_A x_s + \int_{\bar{A}} x_s = Ex_s,$$

a contradiction. Similarly, if $t \in C$, letting $B = \{s = n,\ t > n,$ $E(x_t \mid \mathscr{F}_n) > x_n\}$, $t' = tI_B + sI_{\bar{B}}$ we see that (3.13) holds.

**Lemma 3.3** If $s$ is any s.v. such that for each $n = 1, 2, \ldots$

$$E(x_{n+1} \mid \mathscr{F}_n) \geq x_n \quad \text{on} \quad \{s > n\} \tag{3.14}$$

then $\{x_{\min(s,n)}, \mathscr{F}_n, 1 \leq n < \infty\}$ is a submartingale. If in addition $Ex_s$ exists and

$$\liminf_n \int_{\{s>n\}} x_n^+ = 0, \tag{3.15}$$

then (3.12) holds.

***Proof*** Let $s(n) = \min(s, n)$. Obviously $Ex_{s(n)}^+ \leq \sum_1^n Ex_i^+ < \infty$ and $Ex_{s(n)}$ exists, $n = 1, 2, \ldots$. For all $A \in \mathscr{F}_n$

$$\int_A x_{s(n)} = \int_{A(s\leq n)} x_s + \int_{A(s>n)} x_n$$
$$\leq \int_{A(s\leq n)} x_s + \int_{A(s\geq n+1)} x_{n+1} = \int_A x_{s(n+1)},$$

and hence we conclude that $\{x_{s(n)}, \mathscr{F}_n, 1 \leq n < \infty\}$ is a submartingale.

Now suppose that $Ex_s$ exists. Then for any $A \in \mathscr{F}_n$ $(n = 1, 2, \ldots)$ we have by the submartingale property for each $m = n + 1, n + 2, \ldots$

$$\int_{A(s\geq n)} x_n \leq \int_{A(s\geq n)} x_{s(m)} = \int_{A(n\leq s\leq m)} x_s + \int_{A(s>m)} x_m$$
$$\leq \int_{A(n\leq s\leq m)} x_s + \int_{A(s>m)} x_m^+.$$

Letting $m \to \infty$ along a subsequence $(m')$ for which

$$\int_{(s>m')} x_{m'}^+ \to \liminf_{m\to\infty} \int_{(s>m)} x_m^+ = 0$$

we obtain

$$\int_{A(s\geq n)} x_n \leq \int_{A(s\geq n)} x_s$$

and hence (3.12). The proof remains valid if we assume $Ex_n^+ < \infty$ instead of $E\,|x_n| < \infty$. A similar argument (see problem 1) proves

**Lemma 3.4** If $s, t \in C$ and for each $n = 1, 2, \ldots$

$$E(x_{n+1} \mid \mathscr{F}_n) \leq x_n \quad \text{on} \quad \{s \leq n\} \tag{3.16}$$

and

$$\liminf_n \int_{\{t>n\}} x_n^- = 0, \tag{3.17}$$

then (3.13) holds.

## 5. The Monotone Case

There is one case in which there is a natural candidate for an optimal s.v. This is the *monotone* case, in which the stochastic sequence $\{x_n, \mathscr{F}_n\}_1^\infty$ satisfies certain conditions. Put

$$A_n = \{E(x_{n+1} \mid \mathscr{F}_n) \leq x_n\}, \qquad n = 1, 2, \ldots.$$

We say that we are in the monotone case if

$$A_1 \subset A_2 \subset \cdots; \ \bigcup_1^\infty A_n = \Omega. \tag{3.18}$$

When (3.18) holds, the s.v.

$$s = \text{first } n \geq 1 \quad \text{such that} \quad x_n \geq E(x_{n+1} \mid \mathscr{F}_n) \tag{3.19}$$

deserves special consideration from the point of view of optimality, but before we succumb to the temptation to try to prove that in the monotone case $s$ is always optimal, let us show by two simple examples that sometimes it is not. (These are examples 1 and 3 of the Introduction.)

(a) Let $y_1, y_2, \ldots$ be i.i.d. with $P(y_n = 1) = \frac{1}{2} = P(y_n = -1)$, $\mathscr{F}_n = \mathscr{B}(y_1, \ldots, y_n)$, and set $x_n = 2n/(n+1) \cdot \prod_1^n (y_k + 1)$. Then

$Ex_n = 2n/(n+1)$, $E(x_{n+1} \mid \mathscr{F}_n) = [(n+1)^2/n(n+2)]x_n$, and hence

$$E(x_{n+1} \mid \mathscr{F}_n) \leq x_n \Leftrightarrow x_n = 0 \Rightarrow x_{n+1} = 0 \Rightarrow E(x_{n+2} \mid \mathscr{F}_{n+1}) \leq x_{n+1},$$

so $A_n \subset A_{n+1}$, and $P\left(\bigcup_1^\infty A_n\right) = P(y_n = -1 \text{ for some } n \geq 1) = 1$. Hence we are in the monotone case. Here (3.19) amounts to

$$s = \text{first } n \quad \text{such that} \quad y_n = -1,$$

and $0 = Ex_s \leq Ex_t$ for all $t$. Thus $s$ is the *worst* s.v. It is not hard to show that in this case $V = 2$ and no optimal s.v. exists.

(b) Let $(y_n)$, $(\mathscr{F}_n)$ be as above but put

$$x_n = \min(1, y_1 + \cdots + y_n) - n/(n+1).$$

It is easy to see that $E(x_{n+1} \mid \mathscr{F}_n) < x_n$, $n = 1, 2, \ldots$, so we are trivially in the monotone case and $s$ defined by (3.19) is identically 1, with $Ex_s = -\frac{1}{2}$. Now let $t$ = first $n$ such that $y_1 + \cdots + y_n = 1$. $P(t < \infty) = 1$, so $t$ is a s.v., and since $0 < x_t \leq \frac{1}{2}$, $0 < Ex_t \leq \frac{1}{2}$. Hence $s$ is not optimal; it is easy to see that $t$ is.

The following is the basic result of a positive nature in the monotone case.

**Theorem 3.3** In the monotone case assume that $s$ defined by (3.19) satisfies $P(s<\infty)=1$ and $E(x_s)$ exists. Then if (3.15) holds, $Ex_s \geq Ex_t$ for all $t \in C$ for which (3.17) holds.

***Proof*** The theorem follows directly from Lemmas 3.2, 3.3, and 3.4.

**Corollary** Assume that the conditions of Theorem 3.3, including (3.15), hold, and there exist (i) a random variable $w \geq 0$ with finite expectation and (ii) positive, increasing constants $(c_n)$ such that

$$x_n^- \leq w + c_n \qquad (n = 1, 2, \ldots). \tag{3.20}$$

Then $Ex_s \geq Ex_t > -\infty$ for every s.v. $t$ such that $Ec_t < \infty$.

***Proof*** From (3.20) if $Ec_t < \infty$, $Ex_t^- \leq Ew + Ec_t < \infty$, so $t \in C$; moreover,

$$\int_{\{t>n\}} x_n^- \leq \int_{\{t>n\}} w + \int_{\{t>n\}} c_t \to 0 \ (n \to \infty),$$

so (3.17) holds and Theorem 3.3 applies to prove the assertion of the corollary.

## 6. Applications

(a) Let $y, y_1, y_2, \ldots$ be i.i.d. with $E|y| < \infty$, let

$$\mathscr{F}_n = \mathscr{B}(y_1, \ldots, y_n),$$

$$m_n = \max(y_1, \ldots, y_n), \quad x_n = m_n - c_n \ (n = 1, 2, \ldots),$$

where $(c_n)$ is any strictly increasing sequence of positive constants. Then

$$x_{n+1} - x_n = (y_{n+1} - m_n)^+ - b_n,$$

where we have set $b_n = c_{n+1} - c_n \ (n = 1, 2, \ldots)$. Thus

$$E(x_{n+1} \mid \mathscr{F}_n) \le x_n \Leftrightarrow E((y_{n+1} - m_n)^+ \mid \mathscr{F}_n) \le b_n.$$

Define $\beta_n$ to be the (unique) solution of the equation

$$E(y - \beta_n)^+ = b_n. \tag{3.21}$$

Then

$$E(x_{n+1} \mid \mathscr{F}_n) \le x_n \Leftrightarrow m_n \ge \beta_n. \tag{3.22}$$

Since $m_n \le m_{n+1}$ the first part of (3.18) holds if $\beta_n \ge \beta_{n+1}$, i.e., if $b_{n+1} \ge b_n$.

For the remainder of the current discussion we shall assume that $(b_n)$ is increasing. By (3.22) definition (3.19) becomes

$$s = \text{first } n \ge 1 \quad \text{such that} \quad m_n \ge \beta_n \ (= \infty \text{ if no such } n \text{ exists}).$$

Since $E(y_1 - \beta_1)^+ > 0$ it follows that $p \equiv P(y_1 < \beta_1) < 1$. Hence

$$\begin{aligned} P(s > n) &\le P(y_1 < \beta_1, \ldots, y_n < \beta_n) \\ &\le P(y_1 < \beta_1, \ldots, y_n < \beta_1) = p^n \to 0 \end{aligned}$$

as $n \to \infty$, so $P\{s < \infty\} = 1$ and we are in the monotone case. Moreover,

$$Es^m \le \sum_1^\infty k^m P(s \ge k) \le \sum_1^\infty k^m p^{k-1} < \infty \qquad (m = 1, 2, \ldots),$$

so that $s$ has finite moments of all orders, and by Lemma 3.1 (or Theorem 2.3)

$$Ex_s^+ \le E(y_1^+ + \cdots + y_s^+) = Ey^+ Es < \infty.$$

Since

$$\int_{(s>n)} x_n^+ \leq \beta_1^+ P(s>n) \to 0 \qquad (n \to \infty),$$

(3.15) holds. Condition (3.20) also holds with $w = y_1^-$. It follows from Theorem 3.3 and its corollary that $Ex_s \geq Ex_t$ for every $t \in C$ for which $Ec_t < \infty$.

To go farther and show that $s$ is optimal in $C$ it suffices to show that $Ec_t < \infty$ for every s.v. $t \in C$. To this end we proceed somewhat indirectly. Consider the special case $c_n = c \cdot n$. Then $b_n = c$, $\beta_n = \beta$, where $\beta$ is the unique solution of the equation

$$E(y - \beta)^+ = c;$$

and $s$ = first $n \geq 1$ such that $y_n \geq \beta$. It follows that

$$Ex_s = \sum_1^\infty \left[ P(s \geq n) \int_{(y \geq \beta)} y - cP(s \geq n) \right]$$

$$= \frac{1}{P(y \geq \beta)} \left[ \int_{(y \geq \beta)} y - c \right] = \beta.$$

Let $\beta' > \beta$, and define $S_n = \sum_1^n [(y_k - \beta')^+ - c]$. Then $x_n \leq \beta' + S_n$, so for any s.v. $t$ $x_t \leq \beta' + S_t$. If $Ex_t^- < \infty$, then $ES_t$ exists and by Lemma 3.1

$$ES_t = (Et)(E(y - \beta')^+ - c) < 0.$$

It follows that $Et < \infty$ and that $Ex_t \leq \beta$. Hence in this case $V = \beta$ and $s$ is optimal in $C$.

Suppose now that $(c_n)$ is arbitrary (subject to $0 < c_{n+1} - c_n \uparrow$) and that for some $t \in C$

$$Ec_t = \infty. \tag{3.23}$$

Since $x_n \leq m_n - \frac{1}{2}c_n \leq m_n - \frac{1}{2}nc$ $(n \geq 1)$, where $c = \min(c_1, b_1)$, it follows from the special case $c_n = c \cdot n$ that

$$-\infty < Ex_t \leq E(m_t - \tfrac{1}{2}c_t) \leq E(m_t - \tfrac{1}{2}tc) < \infty.$$

Thus $Ex_t = E(m_t - \frac{1}{2}c_t) - \frac{1}{2}Ec_t = -\infty$ by (3.23). This contradiction shows that $Ec_t < \infty$ for every $t \in C$ and hence that $s$ is optimal in $C$.

***Remark*** It is worth noting that there is contained in the above discussion a proof depending only on Lemma 3.1 and not Theorem 3.3 that in the important special case $c_n = c \cdot n$, $s$ is optimal in $C$.

(b) With the same assumptions as above and $c_n = c \cdot n$, take $x'_n = y_n - c \cdot n$. Then $x'_n \leq x_n$, but $x_s = x'_s$. It follows that $s$ is also optimal for $\{x'_n, \mathscr{F}_n\}_1^\infty$. Note that for this reward sequence we are *not* in the monotone case.

(c) Let $y_1, y_2, \ldots$ be independent and uniformly distributed on the interval $[\theta - \frac{1}{2}, \theta + \frac{1}{2}]$ where $\theta$ is an unknown parameter to be estimated. We assume that if we estimate $\theta$ by $\theta^*$, our loss is $(\theta^* - \theta)^2$ and that we pay a constant amount $c$ to observe each $y_n$, $n = 1, 2, \ldots$. Finally we assume that the unknown $\theta$ has a prior distribution, uniform on $[0, 1]$.

It is easy to see that the posterior distribution of $\theta$ conditional on $y_1$ is uniform on $[u_1, v_1]$, where $[u_1, v_1]$ is the common part of the intervals $[0, 1]$ and $[y_1 - \frac{1}{2}, y_1 + \frac{1}{2}]$, and it follows that the posterior distribution of $\theta$ conditional on $y_1, \ldots, y_n$ is uniform on $[u_n, v_n]$, where $[u_n, v_n]$ is the common part of $[u_{n-1}, v_{n-1}]$ and $[y_n - \frac{1}{2}, y_n + \frac{1}{2}]$. Let $A_n = v_n - u_n$, $n = 1, 2, \ldots$.

A decision rule for the above problem amounts to a stopping rule combined with a terminal decision procedure (see Example 3.1 (h)). It is easily seen by pursuing reasoning analogous to that in Example 3.1 (h) that there is a terminal decision procedure which is uniformly optimal, i.e., optimal independently of the particular s.v. used, that this procedure involves estimating $\theta$ by $(u_n + v_n)/2$ if sampling stops at stage $n$, and that the posterior expected loss is given by $A_n^2/12$. Continuing analogously to the argument of Example 3.1 (h), we see that our problem reduces to the following.

Let $(y_n)$, $(u_n)$, $(v_n)$, $(A_n)$, $\theta$, be as above. Let

$$\mathscr{F}_0 = \{\phi, \Omega\}, x_0 = -A/12,$$

and for $n = 1, 2, \ldots$ let

$$\mathscr{F}_n = \mathscr{B}(y_1, \ldots, y_n), x_n = -A_n^2/12 - c \cdot n.$$

We require an optimal s.v. for $\{x_n, \mathscr{F}_n\}_0^\infty$. Straightforward computations show that

$$E(A_{n+1}^2 \mid A_n = A) = A^2(1 - A/2).$$

It is easily seen that we are in the monotone case and that Theorem 3.2 applies to give an optimal rule.

(d) Given a distribution $F$ on the real line with $\int_{-\infty}^{\infty} x \, dF(x) = 0$, a fixed number of random variables $y_1, \ldots, y_N$ is to be generated according to the following rule. $y_1$ is distributed according to $F$, and at times $n = 2, 3, \ldots, N$ the experimenter may choose an independent new observation (distributed according to $F$) or can repeat the largest of the previous observations. It is desired that the sum of the random variables so generated should have the largest possible expectation. It is obvious that the experimenter need only consider strategies with the property that if at some time $n$ the strategy dictates repeating an observation, then it dictates repeating that observation at time $n + 1$ $(n = 2, 3, \ldots, N - 1)$. In fact, any strategy which for some values $y_1, \ldots, y_{n_0}$ involves repetition at $n_0 + 1, \ldots, n_1$ and a new observation at $n_1 + 1$ is inferior to the strategy which generates a new observation at $n_0 + 1$ and then repeats (with a perhaps larger previous observation) at $n_0 + 2, \ldots, n_1 + 1$, but otherwise resembles the original strategy. The original problem thus reduces to an optimal stopping problem for $\{x_n, \mathscr{F}_n\}_1^N$, where

$$x_n = y_1 + \cdots + y_n + (N - n) \max(y_1, \ldots, y_n),$$

$$\mathscr{F}_n = \mathscr{B}(y_1, \ldots, y_n).$$

($x_n$ is the reward received if $n$ is the *last* index at which a new observation is generated, $n = 1, 2, \ldots, N$.)

A straightforward calculation shows that $E(x_{n+1} - x_n \mid \mathscr{F}_n) = -m_n + (N - n - 1)E((y_{n+1} - m_n)^+ \mid \mathscr{F}_n)$, where we have put $m_n = \max(y_1, \ldots, y_n)$. Thus we are in the monotone case, and $s =$ first $n \geq 1$ such that $m_n \geq \alpha_{N-n}$, where $\alpha_0 = -\infty$ and for $k = 1, 2, \ldots, \alpha_k$ satisfies the equation

$$\alpha_k = (k - 1) \int_{\alpha_k}^{\infty} (y - \alpha_k)\, dF(y). \tag{3.24}$$

It is easy to see that (3.24) has a unique solution with $0 = \alpha_1 < \alpha_2 < \cdots$.

Viewed as a finite problem it is easily seen that the s.v. $s$ defined by (3.19) is in fact the $s^N$ of Theorem 3.2 (see problem 3.3).

## PROBLEMS

**1.** Prove Lemma 3.4. (*Hint:* Give a direct proof or alternatively apply Lemma 3.3 to the supermartingale $\{y_n \equiv E(x_{\max(s,n)} \mid \mathscr{F}_n), \mathscr{F}_n\}_1^\infty$.)

**2.** Show that Example 3.1(e) is a monotone case problem if $y_k \geq 0$, and that Theorem 3.2 applies. What is $s$? Suppose now that we drop the requirement that $y_k$ be non-negative but (assuming $Ey_k \geq 0$) look for a rule which is optimal in the class of rules $t$ with the property that for some positive number $a$, $t$ = first $n$ such that $y_1 + \cdots + y_n \geq a$. This problem then reduces to the non-negative case. It is (heuristically) clear that if there exists an optimal rule, then it must be contained in this subclass of rules. (See problem 5.1.)

**3.** Show that in the monotone case $s^N = \min [N$, first $n$ such that $x_n \geq E(x_{n+1} \mid \mathscr{F}_n)]$.

**4.** Show that $s^N$ is the *minimal* optimal rule in $C^N$, i.e., if $t \in C^N$ and $Ex_t = V^N$, then $t \geq s^N$.

**5.** Suppose that we are in the monotone case and let

$$C^- = \{t: t \in C, \liminf_{n\to\infty} \int_{(t>n)} x_n^- = 0\},$$

$$V^- = \sup_{C^-} Ex_t.$$

Show that if $V^- < \infty$ and there exists an optimal rule in $C^-$ then $s \in C^-$ and $Ex_s = V^-$. (*Hint:* First show that if $t \in C^-$, then $t' = \min(s, t) \in C^-$ and $Ex_{t'} \geq Ex_t$. Then show that if $t \in C^-$ and $P\{t = n < s\} > 0$, $t$ cannot be optimal in $C^-$.) It is easy to see that $V^- \geq \lim_{N\to\infty} V^N$. Is it possible to have strict inequality?

**6.** For the parking problem (Example 3.1(g)), show that we can restrict our consideration to rules of the form

$$t_r = \text{first } k \geq r \quad \text{such that} \quad y_k = 0 \qquad (r \geq Q).$$

Compute $Ex_{t_r}$ and maximize as a function of $r$.

**7.** In the secretary problem suppose that we are interested in maximizing the probability of getting the best girl, but instead of assuming that the girls appear in a random order we suppose that for $n = 0, 1, \ldots, N-1$ the girls appear in the order of the $n$th cyclic permutation of $(N, N-1, \ldots, 1)$ with probability $1/N$. What is $V$? $s^N$? What distribution of the girls has a similar effect if our criterion is to be that of minimizing the expected rank of the girl selected? (*Hint:* The first girl to appear should have probability $\frac{1}{2}$ of being the best of the $N$ girls and probability $\frac{1}{2}$ of being the worst.)

**8.** Apply the fundamental lemmas of Section 3.4 to the sequence $\{\gamma_n^N, \mathscr{F}_n\}_1^N$ to prove Theorem 3.2.

**9.** For the secretary problem of Section 3.3, compute the optimal rule directly by backward induction.

**10.** Find an optimal rule for the secretary problem if the criterion is to maximize the probability of getting one of the two best candidates. What is the limiting value of the maximum probability as $N \to \infty$?

# Chapter 4

# The General Theory

In the finite case, involving $x_1, \ldots, x_N$, we defined

$$\text{(4.1)} \qquad \begin{aligned} \gamma_N^N &= x_N, \\ \gamma_n^N &= \max(x_n, E(\gamma_{n+1}^N \mid \mathscr{F}_n)) \qquad (n = N-1, \ldots, 1). \end{aligned}$$

Theorem 3.2 implies that

$$\text{(4.2)} \qquad \gamma_n^N = \operatorname*{ess\,sup}_{t \in C_n^N} E(x_t \mid \mathscr{F}_n), \quad (n = 1, 2, \ldots, N)$$

(see Section 1.6 for the definition and fundamental property of the essential supremum of a family of r.v.'s) and that

$$\text{(4.3)} \qquad s_n^N = \text{first } i \geq n \quad \text{such that} \quad x_i = \gamma_i^N$$

is optimal in $C_n^N$. In the general optimal stopping problem involving an infinite sequence $x_1, x_2, \ldots$ there is no $x_N$ with which to start the sequence (4.1), and Example 3.1 (a) shows that an optimal rule need not exist. We can, however, *define* $\gamma_n$ for $n = 1, 2, \ldots$ in analogy with (4.2) by

$$\text{(4.2}') \qquad \gamma_n = \operatorname*{ess\,sup}_{C_n} E(x_t \mid \mathscr{F}_n), \qquad (n = 1, 2, \ldots),$$

where $C_n$ is the set of $t \in C$ such that $t \geq n$. We shall then be able to prove that in analogy with (4.1)

(4.1′) $$\gamma_n = \max\,(x_n, E(\gamma_{n+1} \mid \mathscr{F}_n)) \quad (n = 1, 2, \ldots),$$

and to find conditions under which

(4.3′)
$$\sigma_n = \text{first } i \geq n \quad \text{such that} \quad x_i = \gamma_i \; (= \infty \text{ if no such } i \text{ exists})$$

is optimal in $C_n$ (and hence $\sigma = \sigma_1$ is optimal in $C = C_1$). Finally, since (4.2′) defines the $\gamma_n$ non-constructively, we shall show that the finite problem, which has a constructive solution, in some sense approximates the general problem as $N \to \infty$. We shall, for example, find conditions under which $\lim_{N\to\infty} \gamma_n^N = \gamma_n$ $(n = 1, 2, \ldots)$.

## 1. Definitions and Preliminary Lemmas

Our fundamental definitions are the following. Let $C$ denote the class of all s.v.'s $t$ such that $Ex_t^- < \infty$,

$$C_n = \{\max\,(t, n) : t \in C\}, \quad v_n = \sup_{C_n} Ex_t, \quad \gamma_n = \operatorname{ess\,sup}_{C_n} E(x_t \mid \mathscr{F}_n)$$
$$(n = 1, 2, \ldots).$$

It might seem more natural to consider instead of $C_n$ the larger class $C_n^*$ of all s.v.'s $t$ such that $t \geq n$ and $Ex_t$ exists. However, this would yield the same $v_n$ and $\gamma_n$. For if $t \in C_n^*$, define

$$t' = \begin{cases} t & \text{if} \quad E(x_t \mid \mathscr{F}_n) \geq x_n \\ n & \text{if} \quad E(x_t \mid \mathscr{F}_n) < x_n. \end{cases}$$

Putting $A = \{E(x_t \mid \mathscr{F}_n) \geq x_n\}$ we have $Ex_{t'}^- \leq Ex_n^- + \int_A x_t^-$. But $-\infty < \int_A x_n \leq \int_A x_t$, so $\int_A x_t^- < \infty$, and it follows that $t' \in C_n$. Now $E(x_{t'} \mid \mathscr{F}_n) \geq E(x_t \mid \mathscr{F}_n)$ and $Ex_{t'} \geq Ex_t$. It follows that $\gamma_n$ and $v_n$ are unchanged if we replace $C_n$ by $C_n^*$ in their definitions.

The following lemmas will be used in proving the basic general theorems of Section 4.2.

**Lemma 4.1** For each $n = 1, 2, \ldots$ there exists an increasing sequence $(t_k)$ in $C_n$ such that

(4.4) $$x_n \leq E(x_{t_k} \mid \mathscr{F}_n) \uparrow \gamma_n \; (k \to \infty).$$

***Proof*** Choose $(t_k)$ in $C_n$ by Theorem 1.5 such that $t_1 = n$ and $\gamma_n = \sup_k E(x_{t_k} \mid \mathscr{F}_n)$. By Lemmas 4.2 and 4.3 below we can assume that (4.4) holds.

**Lemma 4.2** For any $n = 1, 2, \ldots$ and $t \in C_n$ define

$$t' = \text{first } k \geq n \quad \text{such that} \quad E(x_t \mid \mathscr{F}_k) \leq x_k.$$

Then

(a) $t' \leq t, t' \in C_n,$

(b) $E(x_{t'} \mid \mathscr{F}_n) \geq E(x_t \mid \mathscr{F}_n),$

(c) $t' > j \geq n \Rightarrow E(x_{t'} \mid \mathscr{F}_j) > x_j.$

***Proof*** If $t = j \geq n$, $E(x_t \mid \mathscr{F}_j) = x_j$ and $t' \leq j$; hence $t' \leq t$. Now

$$Ex_{t'}^- = \sum_{k=n}^{\infty} \int_{(t'=k)} x_k^- \leq \sum_{k=n}^{\infty} \int_{(t'=k)} [E(x_t \mid \mathscr{F}_k)]^-$$

$$\leq \sum_{k=n}^{\infty} \int_{(t'=k)} E(x_t^- \mid \mathscr{F}_k) = Ex_t^- < \infty;$$

so $t' \in C_n$. Hence (a) holds. For any $A \in \mathscr{F}_j$ with $j \geq n$

$$\int_{A(t' \geq j)} x_{t'} = \sum_j^{\infty} \int_{A(t'=k)} x_k \geq \sum_j^{\infty} \int_{A(t'=k)} E(x_t \mid \mathscr{F}_k)$$

$$= \int_{A(t' \geq j)} x_t.$$

Putting $j = n$ gives (b). On $\{t' > j\}$ we have

$$E(x_{t'} \mid \mathscr{F}_j) \geq E(x_t \mid \mathscr{F}_j) > x_j,$$

which gives (c).

Any $t \in C_n$ satisfying (c) of the above lemma will be called *n-admissible*. A 1-admissible s.v. will be called *admissible*.

**Lemma 4.3** Let $t_1, t_2 \in C_n$ be $n$-admissible for some fixed $n \geq 1$, and let $\tau = \max(t_1, t_2)$. Then $\tau \in C_n$, is $n$-admissible, and

$$E(x_\tau \mid \mathscr{F}_n) \geq \max_{i=1,2} E(x_{t_i} \mid \mathscr{F}_n).$$

***Proof*** That $\tau \in C_n$ is clear. For any $j \geq n$ and $A \in \mathscr{F}_j$

$$\int_{A(t_1 \geq j)} x_\tau = \sum_j^{\infty} \left[ \int_{A(t_1 = k < \tau)} x_{t_2} + \int_{A(t_1 = k = \tau)} x_k \right]$$

$$\geq \sum_j \left[ \int_{A(t_1 = k < \tau)} x_k + \int_{A(t_1 = k = \tau)} x_k \right] = \int_{A(t_1 \geq j)} x_{t_1}.$$

For $j = n$ this gives $E(x_\tau \mid \mathscr{F}_n) \geq E(x_{t_1} \mid \mathscr{F}_n)$ and by symmetry $E(x_\tau \mid \mathscr{F}_n) \geq \max_{i=1,2} E(x_{t_i} \mid \mathscr{F}_n)$. To prove that $\tau$ is $n$-admissible, we observe that $\{\tau > j\} = \{t_1 > j\} \cup \{t_2 > j\}$ and that on $\{t_i > j\}$, $E(x_\tau \mid \mathscr{F}_j) \geq E(x_{t_i} \mid \mathscr{F}_j) > x_j$ $(i = 1, 2)$.

**Lemma 4.4** If $t \in C$, then for any $n = 1, 2, \ldots$

$$t \geq n \Rightarrow E(x_t \mid \mathscr{F}_n) \leq \gamma_n \quad \text{and} \quad E(x_t^- \mid \mathscr{F}_n) \geq \gamma_n^-.$$

***Proof*** Put $t' = \max(t, n)$. Then $t' \in C_n$ and hence

$$E(x_{t'} \mid \mathscr{F}_n) \leq \gamma_n; E(x_{t'}^- \mid \mathscr{F}_n) \geq [E(x_{t'} \mid \mathscr{F}_n)]^- \geq \gamma_n^-.$$

But $t' = t$ on $\{t \geq n\}$ and the lemma follows.

For each $n = 1, 2, \ldots$ we define the random variable $\sigma_n =$ first $k \geq n$ such that $x_k = \gamma_k$ ($= \infty$ if no such $k$ exists) and write $\sigma_1 = \sigma$. In general $P(\sigma_n < \infty) < 1$, so that $\sigma_n$ is not a s.v.

**Lemma 4.5** If $t \in C_n$, then $t' = \min(t, \sigma_n) \in C_n$ and $E(x_{t'} \mid \mathscr{F}_n) \geq E(x_t \mid \mathscr{F}_n)$.

***Proof*** Writing $\sigma_n = \sigma$ and using Lemma 4.4, for any $A \in \mathscr{F}_n$

$$\text{(4.5)} \qquad \int_{A(\sigma<t)} x_\sigma^- = \sum_{k=n}^{\infty} \int_{A(\sigma=k<t)} x_k^- = \sum_{k=n}^{\infty} \int_{A(\sigma=k<t)} \gamma_k^-$$

$$\leq \sum_{k=n}^{\infty} \int_{A(\sigma=k<t)} x_t^- = \int_{A(\sigma<t)} x_t^-.$$

Hence $Ex_{t'}^- = \int_{(t\leq\sigma)} x_t^- + \int_{(\sigma<t)} x_\sigma^- \leq Ex_t^- < \infty$. Now replacing $(\cdot)^-$ by $(\cdot)$ in (4.5) and reversing the inequalities, we have $\int_A x_{t'} = \int_{A(t<\sigma)} x_t + \int_{A(\sigma<t)} x_\sigma \geq \int_A x_t$.

**Lemma 4.6** Suppose that $V < \infty$ and that $(t_k)$ is an increasing sequence of s.v.'s in $C$ such that $Ex_{t_k} \uparrow V$ as $k \to \infty$. Then $P(\lim t_k \geq \sigma) = 1$.

***Proof*** For any $n = 1, 2, \ldots$ let $(s_k)$ be as in Lemma 4.1. Then by the monotone convergence theorem $V \geq Ex_{s_k} = E(E(x_{s_k} \mid \mathscr{F}_n)) \uparrow E\gamma_n$, and it follows that $E\gamma_n < \infty$ $(n = 1, 2, \ldots)$. Put $t = \lim t_k$ and suppose that for some $i$, $A = \{t = i < \sigma\}$ has positive probability. There exists an $\varepsilon > 0$ such that

$$\int_A \gamma_i - 3\varepsilon \geq \int_A x_i.$$

For each $k \geq 1$ let $B_k = \{t_k = i < \sigma\}$; then $I_{B_k} \to I_A$, and by the dominated convergence theorem there exists a $k_0$ such that for all $k \geq k_0$

$$(4.6) \qquad \int_{B_k} x_i \leq \int_{B_k} \gamma_i - 2\varepsilon.$$

By Lemma 4.1 there exists for each $k \geq k_0$ a $t'_k \in C_i$ such that

$$(4.6') \qquad \int_{B_k} x_{t'_k} \geq \int_{B_k} \gamma_i - \varepsilon.$$

Letting $\tau_k = t'_k I_{B_k} + t_k I_{\bar{B}_k}$, it is easily seen that $\tau_k \in C$ and from (4.6) and (4.6') we have for all $k \geq k_0$

$$Ex_{\tau_k} = \int_{B_k} x_{t'_k} + \int_{\bar{B}_k} x_{t_k} \geq \int_{B_k} x_i + \int_{\bar{B}_k} x_{t_k} + \varepsilon = Ex_{t_k} + \varepsilon.$$

Hence $\sup_k Ex_{\tau_k} \geq V + \varepsilon$, a contradiction.

## 2. Some General Theorems

We can now prove that (4.1′) always holds.

**Theorem 4.1**

$$\text{(a)} \quad \gamma_n = \max(x_n, E(\gamma_{n+1} \mid \mathscr{F}_n)),$$

$$\text{(b)} \quad v_n = E\gamma_n \qquad (n = 1, 2, \ldots).$$

***Proof*** Let $n = 1, 2, \ldots$ and $t \in C_n$ be arbitrary, and let $B = \{t = n\}$. By Lemma 4.4 $E(x_t \mid \mathscr{F}_{n+1}) \leq \gamma_{n+1}$ on $\bar{B}$ and hence $E(x_t \mid \mathscr{F}_n) \leq E(\gamma_{n+1} \mid \mathscr{F}_n)$ on $\bar{B}$. Thus

$$\begin{aligned} E(x_t \mid \mathscr{F}_n) &= I_B x_n + I_{\bar{B}} E(x_t \mid \mathscr{F}_n) \\ &\leq \max(x_n, E(\gamma_{n+1} \mid \mathscr{F}_n)); \end{aligned}$$

therefore

$$\gamma_n \leq \max(x_n, E(\gamma_{n+1} \mid \mathscr{F}_n)).$$

To complete the proof let $(t_k)$ in $C_{n+1}$ be as in Lemma 4.1. Then by the monotone convergence theorem for conditional expectations

$$\gamma_n \geq E(x_{t_k} \mid \mathscr{F}_n) = E[E(x_{t_k} \mid \mathscr{F}_{n+1}) \mid \mathscr{F}_n] \uparrow E(\gamma_{n+1} \mid \mathscr{F}_n) \quad \text{as} \quad k \to \infty.$$

Since the inequality $\gamma_n \geq x_n$ is trivially satisfied, the proof is complete.

**Theorem 4.2** (a) If $\sigma \in C$ and is admissible, it is optimal. (b) If $V < \infty$ and an optimal s.v. exists, then $\sigma \in C$ and is optimal and admissible, and

$$\sigma \geq n \Rightarrow E(x_\sigma \mid \mathscr{F}_n) = \gamma_n \qquad (n = 1, 2, \ldots). \tag{4.7}$$

***Proof***

(a) By Lemma 4.4 condition (3.13) is satisfied with $s = \sigma$, and by hypothesis condition (3.12) is satisfied. Lemma 3.2 completes the proof.

(b) Lemmas 4.5 and 4.6 imply the optimality of $\sigma$. To show that $\sigma$ is admissible it suffices to verify (4.7). For any $n = 1, 2, \ldots$ let $A = \{\sigma > n, E(x_\sigma \mid \mathscr{F}_n) < \gamma_n\}$. If $P(A) > 0$, then $\int_A x_\sigma < \int_A \gamma_n$, since $E\gamma_n \leq V < \infty$. By Lemma 4.1 there exists $t \in C_n$ such that $\int_A x_t > \int_A x_\sigma$. Let $\tau = tI_A + \sigma I_{\bar{A}}$. It is easy to see that $\tau \in C$ and $Ex_\tau > Ex_\sigma$, a contradiction.

Theorem 4.2 is perhaps the main result of the general theory. However, the random variables $\gamma_n$ and the constants $v_n$ are in general impossible to compute directly from the definitions at the beginning of Section 4.1. In many cases a constructive scheme for obtaining the $\gamma_n$ and $v_n$ is provided by the method of truncation. To apply this method we define for any $N \geq 1$ and $n = 1, 2, \ldots, N$ the expressions

$$C_n^N = \{\min(t, N) : t \in C_n\}, \quad v_n^N = \sup_{C_n^N} Ex_t,$$
$$\gamma_n^N = \operatorname*{ess\,sup}_{C_n^N} E(x_t \mid \mathscr{F}_n).$$

Then

$$-\infty < Ex_n = v_n^n \leq v_n^{n+1} \leq \cdots \leq v_n,$$
$$x_n = \gamma_n^n \leq \gamma_n^{n+1} \leq \cdots \leq \gamma_n,$$

so that we can define

$$v_n' = \lim_{N \to \infty} v_n^N, \quad \gamma_n' = \lim_{N \to \infty} \gamma_n^N \tag{4.8}$$

and have

$$-\infty < Ex_n \leq v_n' \leq v_n, \quad x_n \leq \gamma_n' \leq \gamma_n.$$

By the argument of Theorem 4.1 applied to the *finite* sequence $\{x_n, \mathscr{F}_n\}_1^N$, we have

$$\gamma_N^N = x_N,$$
$$\gamma_n^N = \max(x_n, E(\gamma_{n+1}^N \mid \mathscr{F}_n)) \qquad (n = N - 1, \ldots, 1),$$

and $E\gamma_n^N = v_n^N$, so that $(\gamma_n^N)$ and $(v_n^N)$ are computable by recursion, and our present definition of $(\gamma_n^N)$ and $(v_n^N)$ is consistent with that of Chapter 3. By the monotone convergence theorem

$$E\gamma'_n = v'_n$$

and

$$\gamma'_n = \max(x_n, E(\gamma'_{n+1} \mid \mathscr{F}_n)) \qquad (n = 1, 2, \ldots). \tag{4.9}$$

*Hence* $(\gamma'_n)$ *satisfies the same recursion relation as does* $(\gamma_n)$, but it is *not* in general the case that $(\gamma'_n) = (\gamma_n)$. (Suppose, for example, that $y_1, y_2, \ldots$ are i.i.d., $P\{y_1 = 1\} = \frac{1}{2} = P\{y_1 = -1\}$. Let $x_n = \sum_1^n y_k$, $\mathscr{F}_n = \mathscr{B}(y_1, \ldots, y_n)$ $(n \geq 1)$. Since $\{x_n, \mathscr{F}_n, 1 \leq n < \infty\}$ is a martingale, $E(x_t \mid \mathscr{F}_n) = x_n$ on $\{t \geq n\}$ for every *bounded* s.v. $t$ by Theorem 2.2, and hence $\gamma'_n = x_n$ $(n \geq 1)$. But $P\{\limsup x_n = +\infty\} = 1$, so $\gamma_n = +\infty$ for all $n$.) However

**Theorem 4.3** For any $n = 1, 2, \ldots$ if

$$\liminf_{N \to \infty} \int_{\{t > N\}} (\gamma'_N)^- = 0 \tag{4.10}$$

for every $t \in C_n$, then $\gamma'_n = \gamma_n$ and $v'_n = v_n$. In particular, $v'_1 = V$.

***Proof*** The equations (4.9) imply that $\{\gamma'_n, \mathscr{F}_n, 1 \leq n < \infty\}$ is a supermartingale. By Lemma 3.3 (or Theorem 2.2)

$$\gamma'_n \geq E(\gamma'_t \mid \mathscr{F}_n)$$

for every $t \in C_n$. Since $\gamma'_k \geq x_k$ $(k = 1, 2, \ldots)$ it follows that

$$\gamma'_n \geq E(x_t \mid \mathscr{F}_n)$$

for every $t \in C_n$ and hence

$$\gamma'_n \geq \gamma_n.$$

Since the reverse inequality $\gamma'_n \leq \gamma_n$ holds in general, the proof is complete.

***Remark*** By Lemma 4.5, Theorem 4.3 remains true if we assume only that (4.10) holds for those $t \in C_n$ such that $P\{t \leq \sigma_n\} = 1$. This observation will be used in Section 5.7.

It is desirable, of course, to state easily verified conditions on the sequence $\{x_n, \mathscr{F}_n\}_1^\infty$ which imply (4.10) and hence the conclusion of Theorem 4.3. To this end it suffices to show that

$$\int_{(t>n)} x_n^- \to 0 \ (t \in C)$$

and hence to exhibit sequences $(z_{1,n}), \ldots, (z_{K,n})$ of non-negative r.v.'s for which

$$x_n^- \leq \sum_{k=1}^{K} z_{k,n} \qquad (n \geq 1)$$

and

$$\lim_{n\to\infty} \int_{(t>n)} z_{k,n} = 0 \qquad (k = 1, \ldots, K; t \in C).$$

Two classes of sequences $(z_n)$ useful in applications are

$(z_n)$ uniformly integrable

and

$(z_n)$ increasing with $Ez_t < \infty$ for all $t \in C$.

The following theorem and its corollary are motivated by these considerations and by the so-called "statistical model," in which $-x_n$ is the sum of two non-negative $\mathscr{F}_n$-measurable r.v.'s, a loss due to making an incorrect terminal decision and a loss due to the cost of sampling, which increases to $+\infty$ with the number of observations. (See Sections 3.1(h) and 3.6(c).)

**Theorem 4.4** Suppose that $x_n = x_n' - x_n''$, where $x_n'$ and $x_n''$ are $\mathscr{F}_n$-measurable $(n \geq 1)$, and that

(a) $((x_n')^-)$ is uniformly integrable;
(b) there exists a non-negative, increasing stochastic sequence $\{z_n, \mathscr{F}_n\}_1^\infty$ for which $x_n'' \leq z_n$ $(n \geq 1)$ and $Ez_t < \infty$ for all $t \in C$.

Then $\gamma_n' = \gamma_n$ for all $n = 1, 2, \ldots$.

***Proof*** By Theorem 4.3 it suffices to show that

$$\int_{(t>n)} x_n^- \to 0 \qquad (n \to \infty)$$

for each $t \in C$. But $x_n^- \leq (x_n')^- + (x_n'')^+ \leq (x_n')^- + z_n$ and hence by (a), (b), and Theorem 1.3

$$\int_{(t>n)} x_n^- \leq \int_{(t>n)} (x_n')^- + \int_{(t>n)} z_n \leq \int_{(t>n)} (x_n')^- + \int_{(t>n)} z_t \to 0 \quad (n \to \infty)$$

**Corollary** If $x_n = \hat{x}_n - \hat{\hat{x}}_n = x_n^* - x_n^{**}$, where all components are $\mathscr{F}_n$-measurable $(n \geq 1)$; and if

(a) $((x_n^*)^-)$ is uniformly integrable,
(b) $E(\sup_n \hat{x}_n^+) < \infty$,
(c) $0 \leq \hat{\hat{x}}_1 \leq \hat{\hat{x}}_2 \leq \cdots$,

and

(d) for some $C > 0$ $x_n^{**} \leq C\hat{\hat{x}}_n$ $(n \geq 1)$,

$$\text{then } \gamma_n' = \gamma_n \text{ for all } n = 1, 2, \ldots .$$

***Proof*** Let $x_n' = x_n^*$, $x_n'' = x_n^{**}$, $z_n = C\hat{\hat{x}}_n$. By (b) $Ez_t < \infty$ for all $t \in C$, and the corollary follows at once.

***Remark*** Theorem 4.4 remains true if we no longer insist that $z_n \uparrow$ but assume that $\{z_n, \mathscr{F}_n, 1 \leq n < \infty\}$ is a $C$-regular submartingale (defined in Section 4.4). Condition (c) of the corollary may be similarly weakened.

The statistical model also suggests conditions sufficient to guarantee that $\sigma$ is in fact optimal.

**Theorem 4.5** Suppose $E(\sup x_n^+) < \infty$. If $P(\sigma < \infty) = 1$, then $\sigma$ is optimal. In order that $P(\sigma < \infty) = 1$ it suffices that $\lim x_n = -\infty$.

***Proof*** Choose $t_i$ admissible and increasing such that $P\{t_i \leq \sigma\} = 1$ for all $i$ and $Ex_{t_i} \uparrow V$ (by Lemmas 4.2, 4.3, and 4.5 this can always be done). By Lemma 4.6, $\sigma = \lim t_i$. By Fatou's lemma, $V$ equals

$$\lim_{i\to\infty} Ex_{t_i} \leq E(\limsup_{i\to\infty} x_{t_i}) = \int_{(\sigma<\infty)} x_\sigma + \int_{(\sigma=\infty)} (\limsup_{n\to\infty} x_n).$$

If $P(\sigma < \infty) = 1$, then $Ex_\sigma \geq V$ and $\sigma$ is optimal. If $x_n \to -\infty$, then we must have $P(\sigma < \infty) = 1$.

We remark that if in addition to the conditions of the corollary above, $\hat{\hat{x}}_n \to \infty$, then by Theorem 4.5 $\sigma$ is optimal.

## 3. Applications

(a) Theorem 4.5 applies at once to the problems of Example 3.1(h) and 3.6(c) to show that $\sigma$ is optimal. In each case an easy application of Theorem 4.4 provides in principle a method for computing $\sigma$.

(b) Suppose that we are in the monotone case, so that

$$E(x_{n+2} \mid \mathscr{F}_{n+1}) \leq x_{n+1} \quad \text{on} \quad \{E(x_{n+1} \mid \mathscr{F}_n) \leq x_n\}$$

(see Section 3.5). By Problem 3.3

$$s^N = \min (N, \text{first } n \text{ such that } x_n \geq E(x_{n+1} \mid \mathscr{F}_n)),$$

and hence $s$ defined by (3.19) is $\lim_{N\to\infty} s^N$. Lemma 4.6 (see also Problem 4.3) says that in order that $s = \sigma$ it suffices that $V^N \to V < \infty$; Theorem 4.4 provides a method for verifying that $V^N \to V$; and Theorem 4.5 tells us conditions under which $\sigma$ is optimal. We now apply these tools to the problems of Section 3.6(a), (b) (not all of which are monotone case problems).

We use the following lemma, the proof of which will be deferred.

**Lemma 4.7** Let $w, w_1, w_2, \ldots$ be identically distributed, non-negative random variables, and for any $\alpha > 0$ set

$$z = \sup_n (\max (w_1, \ldots, w_n) - n^\alpha).$$

Then

(i) $\quad Ew^{\alpha^{-1}} < \infty$ implies that $P\{z < \infty\} = 1$;

(ii) for any $\beta > 0$, if $Ew^{\alpha^{-1}+\beta} < \infty$, then $E(z^+)^\beta < \infty$.

Now let $y, y_1, y_2, \ldots$ be identically distributed, let $\alpha$ be a positive constant, and let

$$x_n = \max (y_1, \ldots, y_n) - n^\alpha, \tilde{x}_n = y_n - n^\alpha \qquad (n = 1, 2, \ldots).$$

If $E|y| < \infty$ and $E(y^+)^{1+\alpha^{-1}} < \infty$ there exist s.v.'s $\sigma$ and $\tilde{\sigma}$ such that $V = Ex_\sigma$, $\tilde{V} = E\tilde{x}_{\tilde{\sigma}}$, and $\sigma = \lim s^N$, $\tilde{\sigma} = \lim \tilde{s}^N$. If, moreover, the $y_n$ are independent, we have for $\alpha \geq 1$,

$$\sigma = \text{first } n \quad \text{such that} \quad \max (y_1, \ldots, y_n) \geq \beta_n$$

where $\beta_n$ is defined by $E[(y - \beta_n)^+] = (n + 1)^\alpha - n^\alpha$.

***Proof*** Since Lemma 4.7 implies that

$$E(\sup_n [\max(y_1, \ldots, y_n) - n^\alpha/2]) < \infty,$$

by putting $\hat{x}_n = \max(y_1, \ldots, y_n) - \frac{1}{2}n^\alpha, x_n^* = \max(y_1, \ldots, y_n)$, we see at once that the corollary to Theorem 4.4 applies to $\{x_n, \mathscr{F}_n\}_1^\infty$, and hence $\sigma = \lim_{N\to\infty} s^N$. By Theorem 4.5 $\sigma$ is optimal in $C$. A similar argument applies to $\{\tilde{x}_n, \mathscr{F}_n\}_1^\infty$. If the $y_n$ are independent and $\alpha \geq 1$, then the optimal stopping problem for $\{x_n, \mathscr{F}_n\}_1^\infty$ is a monotone case problem, and $\sigma$ is the s.v. $s$ defined by (3.19).

***Proof of Lemma 4.7*** It is easy to see that $z = \sup_n (w_n - n^\alpha)$. Hence by the Borel-Cantelli lemma (Section 2.4(d))

(i)
$$\begin{aligned} Ew^{\alpha-1} < \infty &\Rightarrow \sum_1^\infty P\{2w > n^\alpha\} < \infty \\ &\Rightarrow \sum_1^\infty P\{2w_n > n^\alpha\} < \infty \\ &\Rightarrow P\left(\bigcap_{n=1}^\infty \bigcup_{k=n}^\infty \left\{w_k > \frac{k^\alpha}{2}\right\}\right) = 0 \\ &\Rightarrow P\{z < \infty, w_n - n^\alpha \to -\infty\} = 1. \end{aligned}$$

(ii) Assume that $Ew^{\alpha-1+\beta} < \infty$. To show that $E(z^+)^\beta < \infty$, it suffices to show that

$$\int_0^\infty u^{\beta-1} P\{z > u\}\, du < \infty.$$

Now for any $u > 0$

$$\begin{aligned} P\{z > u\} &\leq \sum_{n=1}^\infty P\{w > u + n^\alpha\} \\ &= \sum_{n=1}^\infty \sum_{k=n}^\infty P\{k^\alpha < w - u \leq (k+1)^\alpha\} \\ &= \sum_{k=1}^\infty kP\{k^\alpha < w - u \leq (k+1)^\alpha\} \leq E((w-u)^+)^{\alpha^{-1}}. \end{aligned}$$

Putting $F(x) = P\{w \leq x\}$ and $\gamma = \alpha^{-1}$, we have

$$\begin{aligned} E((w-u)^+)^{\alpha^{-1}} &= \int_u^\infty (x-u)^\gamma\, dF(x) \\ &= \gamma \int_u^\infty (x-u)^{\gamma-1}(1 - F(x))\, dx. \end{aligned}$$

Hence

$$\int_0^\infty u^{\beta-1}P\{z > u\}\,du \le \gamma \int_0^\infty u^{\beta-1} \int_u^\infty (x-u)^{\gamma-1}(1 - F(x))\,dx\,du$$
$$= \gamma \int_0^\infty \left(\int_0^x u^{\beta-1}(x-u)^{\gamma-1}\,du\right)(1 - F(x))\,dx$$
$$\le \text{const} \int_0^\infty x^{\beta+\gamma-1}(1 - F(x))\,dx < \infty.$$

(c) Suppose that we are observing the output of a production process in the form of i.i.d. random variables $y_1, y_2, \ldots$. A change occurs in the process at some random time $\theta$. The problem is one of detecting the occurrence of a change.

Simplifying and formalizing we assume that $\theta$ is a non-negative integer-valued random variable such that

$$P\{\theta = 0\} = \pi$$
$$P\{\theta = k \mid \theta > 0\} = r_k, \qquad k = 1, 2, \ldots, \left(\sum_1^\infty r_k = 1\right).$$

We observe $y_1, \ldots, y_{\theta-1}, y'_\theta, y'_{\theta+1}, \ldots,$ where $y_1, y_2, \ldots, (y'_1, y'_2, \ldots)$ are i.i.d. according to a known distribution $F_0(F_1)$. If we stop the process at time $n$, we lose

$$\begin{aligned} &c \text{ (a fixed inspection cost)} && \text{if} \quad \theta > n, \\ &n - \theta && \text{if} \quad \theta \le n. \end{aligned}$$

We want to minimize our expected loss.

Let $\mathscr{F}_n$ be the $\sigma$-algebra generated by the first $n$ random variables we observe, but as in Example 3.1 (e) our loss if we stop at stage $n$ is not, as phrased above, $\mathscr{F}_n$-measurable. We remedy this situation by setting

$$x_n = -c(1 - \pi_n) - \sum_{i=0}^{n-1} (n - i)p_i^n \qquad (n = 0, 1, 2, \ldots),$$

where we have put

$$p_i^n = P(\theta = i \mid \mathscr{F}_n) \qquad (i, n \ge 0),$$
$$\pi_n = P(\theta \le n \mid \mathscr{F}_n) \qquad (n \ge 0).$$

We now show that the conditions of Theorem 4.3 and 4.5 are satisfied and hence that $\sigma$ is optimal and can be computed by backward induction and passing to the limit. Obviously $E(\sup x_n^+) < \infty$.

Putting $\mathscr{F}_\infty = \mathscr{B}\left(\bigcup_1^\infty \mathscr{F}_n\right)$, we have by Theorem 1.4 for each $n \geq N = 1, 2, \ldots$,

$$\sum_0^n (n - i)p_i^n \geq (n - N)P\{\theta \leq N \mid \mathscr{F}_n\} \to \infty$$
$$\text{as} \quad n \to \infty \quad \text{on} \quad \{P(\theta \leq N \mid \mathscr{F}_\infty) > 0\}.$$

But $\bigcup_N \{P(\theta \leq N \mid \mathscr{F}_\infty) > 0\} = \Omega$ and hence $\sum_{i=0}^n (n - i)p_i^n \to \infty$ $(n \to \infty)$ with probability one. Hence $\sigma$ is optimal in $C$ by Theorem 4.5. Since $c(1 - \pi_n) \leq c$ $(n \geq 1)$, to show that $(\gamma_n') = (\gamma_n)$ it suffices by Theorem 4.3 to show $\lim_{n\to\infty} \int_{(t>n)} \left(\sum_{i=0}^n (n - i)p_i^n\right) = 0$ $(t \in C)$. However, since $\{p_i^n, \mathscr{F}_n\}_i^\infty$ is a bounded martingale, we have by Lemma 3.3 (Theorem 2.2) for any $t \in C$

$$\int_{(t>n)} \sum_{i=0}^n (n - i)p_i^n = \sum_{i=0}^n (n - i) \int_{(t>n)} p_i^t$$
$$\leq \int_{(t>n)} \sum_{i=0}^t (t - i)p_i^t \to 0$$
$$(n \to \infty).$$

(d) The idea underlying Theorem 4.4 is that to show that $V^N \to V$, it certainly suffices to show that $\liminf_{N\to\infty} Ex_{\min(t,N)} \geq Ex_t$ for every $t \in C$. That this is by no means necessary is the content of the following example, in which $V = \lim_{N\to\infty} V^N = 1$ and $\sigma$ is optimal, but $Ex_{\min(\sigma,N)} = 0$ for every $N$.

Let $x_1, x_2, \ldots$ be independent with

$$P\left(x_n = \begin{cases} 1 \\ 0 \\ -2n \end{cases}\right) = \begin{cases} \dfrac{1}{n+1} \\ \dfrac{n}{2(n+1)}, \\ \dfrac{n}{2(n+1)} \end{cases} \qquad \mathscr{F}_n = \mathscr{B}(x_1, \ldots, x_n).$$

Clearly, $\sigma$ ( $=$ first $n$ such that $x_n = 1$) is optimal, and $V = 1$. Moreover,

$$Ex_{\min(\sigma,N)} = P(\sigma \leq N) - NP(\sigma > N)$$
$$= 1 - \frac{1}{N+1} - \frac{N}{N+1} = 0$$

for every $N \geq 1$. Now given $n \geq 1$ define

$$t = \begin{cases} \text{first } 1 \leq k \leq n \quad \text{such that} \quad x_k = 1 \quad \text{or if no such } k \text{ exists} \\ \text{first } n+1 \leq k \leq n^2 - 1 \quad \text{such that} \quad x_k \neq -2k \\ \qquad\qquad\qquad\qquad\qquad\qquad \text{or if no such } k \text{ exists} \\ n^2. \end{cases}$$

Then $t \leq n^2$, so $Ex_t \leq V^{n^2}$. However,

$$P\{x_t = 1\} \geq P\{\sigma \leq n\} = \frac{n}{n+1},$$

$$P\{x_t = -2n^2\} = \frac{1}{n+1} \cdot \frac{n+1}{2(n+2)} \cdot \frac{n+2}{2(n+3)} \cdot \cdots$$
$$\cdot \frac{n^2}{2(n^2+1)} = \frac{1}{2^{n^2-n}(n^2+1)},$$

so $Ex_t \geq \dfrac{n}{n+1} - \dfrac{2n^2}{2^{n^2-n}(n^2+1)} \to 1$ as $n \to \infty$.

## 4. A Martingale Characterization of $(\gamma_n)$ and $(\gamma'_n)$

**Definition** A supermartingale $\{y_n, \mathscr{F}_n, 1 \leq n < \infty\}$ is said to be regular with respect to a class of s.v.'s $D$ if for each $t \in D$, $Ey_t$ exists and $E(y_t \mid \mathscr{F}_n) \leq y_n$ on $\{t \geq n\}$ $(n \geq 1)$.

In this section we indicate briefly how a systematic exploitation of this concept provides an alternative approach to many of the results of the present chapter. If, for example, $\{\beta_n, \mathscr{F}_n\}_1^\infty$ is a $C$-regular supermartingale such that $\beta_n \geq x_n$ $(n = 1, 2, \ldots)$, then $\beta_n \geq E(\beta_t \mid \mathscr{F}_n) \geq E(x_t \mid \mathscr{F}_n)$ for all $t \in C_n$, and hence $\beta_n \geq \gamma_n$ $(n = 1, 2, \ldots)$. Since $\{\gamma'_n, \mathscr{F}_n\}_1^\infty$ is a supermartingale by (4.9), Theorem 4.4 may be interpreted as establishing under certain conditions the $C$-regularity of $\{\gamma'_n, \mathscr{F}_n\}_1^\infty$.

**Definition** If $Y = \{y_n, \mathscr{F}_n\}_1^\infty$ and $Z = \{z_n, \mathscr{F}_n\}_1^\infty$ are two stochastic sequences on $(\Omega, \mathscr{F}, P)$, we say that $Y$ *dominates* $Z$ if $y_n \geq z_n$ for each $n = 1, 2, \ldots$. If $Y$ and $Z$ are stochastic sequences and there exists a class $\mathscr{K}$ of stochastic sequences such that

(a) $\qquad Y \in \mathscr{K},$

(b) $\qquad Y$ dominates $Z,$

and

(c) $\qquad$ every $Y' \in \mathscr{K}$ dominating $Z$ also dominates $Y,$

then we say that $Y$ is the *minimal* element of $\mathscr{K}$ dominating Z.

**Theorem 4.6** (a) $\{\gamma_n', \mathscr{F}_n\}_1^\infty$ is the minimal supermartingale dominating $\{x_n, \mathscr{F}_n\}_1^\infty$. (b) $\{\gamma_n, \mathscr{F}_n\}_1^\infty$ is the minimal $C$-regular supermartingale dominating $\{x_n, \mathscr{F}_n\}_1^\infty$.

***Proof*** That $\{\gamma_n', \mathscr{F}_n\}_1^\infty$ is a supermartingale dominating $\{x_n, \mathscr{F}_n\}_1^\infty$ is clear from (4.9). If $\{\beta_n, \mathscr{F}_n\}_1^\infty$ is any other, then

$$\beta_N \geq x_N = \gamma_N^N \qquad (N = 1, 2, \ldots)$$

and by backward induction

$$\beta_n \geq \gamma_n^N \qquad (N = 1, 2, \ldots; n = 1, 2, \ldots, N).$$

Letting $N \to \infty$, we have

$$\beta_n \geq \gamma_n' \qquad (n = 1, 2, \ldots).$$

(b) It suffices to establish the $C$-regularity of $\{\gamma_n, \mathscr{F}_n\}_1^\infty$; the minimality is clear from the remarks at the beginning of this section. Assume that $E(\sup x_n^+) < \infty$. (This condition will be removed by Lemma 4.12.) If $\{\gamma_n, \mathscr{F}_n\}_1^\infty$ is not $C$-regular there exist an integer $n$, an $\varepsilon > 0$, and a set $A \in \mathscr{F}_n$ such that for some $t \in C$

$$\int_{A(t \geq n)} \gamma_n + \varepsilon \leq \int_{A(t \geq n)} \gamma_t.$$

By Lemma 4.1 for each $k = n, n+1, \ldots,$ there exists a s.v. $t_k \in C_k$ such that

$$\int_{A(t=k)} \gamma_k - \frac{\varepsilon}{2^{k+1}} \leq \int_{A(t=k)} x_{t_k}.$$

Putting $t^* = \sum_n^\infty t_k I_{A(t=k)} + n I_{\Omega - [A(t \geq n)]}$, we see that $t^*$ is a s.v. and

$$\int_{A(t \geq n)} E(x_{t^*} \mid \mathscr{F}_n) = \sum_{k=n}^\infty \int_{A(t=k)} x_{t_k} \geq \sum_{k=n}^\infty \int_{A(t=k)} \gamma_k - \frac{\varepsilon}{2}$$

$$= \int_{A(t \geq n)} \gamma_t - \frac{\varepsilon}{2} \geq \int_{A(t \geq n)} \gamma_n + \frac{\varepsilon}{2}.$$

Thus $t^* \in C_n$ and $P\{E(x_{t^*} \mid \mathscr{F}_n) > \gamma_n\} > 0$, a contradiction.

As an example of an application of Theorem 4.6 we provide the following alternative proof of Lemma 4.6. Let $k_0$ be such that (4.6) holds for all $k \geq k_0$. Since

$$\int_{\bar{B}_k} x_{t_k} \leq \int_{\bar{B}_k} \gamma_{t_k},$$

we have by the $C$-regularity of $\{\gamma_n, \mathscr{F}_n\}_1^\infty$ that

$$Ex_{t_k} \le E\gamma_{t_k} - 2\varepsilon \le E\gamma_1 - 2\varepsilon = V - 2\varepsilon,$$

a contradiction to the hypothesis of the lemma.

## 5. Extended Stopping Variables. The Triple Limit Theorem

Although some problems "arising in practice" may be sufficiently well-behaved to permit computing the $\gamma_n$ by computing the $\gamma_n^N$ and passing to the limit, nevertheless, the central role assigned to the sequence $(\gamma_n)$ in our theory makes it desirable to have computational techniques that are always valid. The present section provides such techniques; and although it is unlikely that one would find it desirable to carry out the computations, the *qualitative* properties of the sequence $(\gamma_n)$ which may be inferred from these results are frequently quite useful.

We shall use the following notation throughout the remainder of this section (except for Theorem 4.9). Let $-\infty \le a < 0 < b \le \infty$,

$$x_n(a, b) = \begin{cases} b & \text{if } x_n > b \\ x_n & \text{if } a \le x_n \le b \\ a & \text{if } x_n < a \end{cases}$$

$$x_n(b) = x_n(-\infty, b)$$

$$x_n(a) = x_n(a, +\infty).$$

Let $\gamma_n(a, b)$ ($\gamma_n^N(b)$, etc.) denote the $\gamma_n$($\gamma_n^N$, etc.) associated with $\{x_n(a, b), \mathscr{F}_n\}_1^\infty$ ($\{x_n(b), \mathscr{F}_n\}_1^N$, etc.).

Since by Theorem 4.3, for all $a < 0$ $\lim_{N\to\infty} \gamma_n^N(a) = \gamma_n(a)$, $\lim_{N\to\infty} v_n^N(a) = v_n(a)$ $(n = 1, 2, \ldots)$, it seems natural to ask if

$$\lim_{a\to-\infty} \gamma_n(a) = \gamma_n, \quad \lim_{a\to-\infty} v_n(a) = v_n \qquad (n = 1, 2, \ldots). \tag{4.11}$$

However, without further assumptions (4.11) is not true (see Section 4.6(a)). The reason for this, loosely speaking, is that there may exist rules $t_k$ such that

$$\sup_k Ex_{t_k}^+ = \sup_k Ex_{t_k}^- = \infty, \quad \text{but} \quad Ex_{t_k} = Ex_{t_k}^+ - Ex_{t_k}^-$$

remains bounded. Thus it may happen that $V < \infty$, but $V(a) = \infty$ for all $a > -\infty$. To avoid this difficulty we proceed to prove (4.11) under the hypothesis that $E(\sup_n x_n^+) < \infty$. Then we show that

$$\lim_{b\to\infty} \gamma_n(b) = \gamma_n,$$

which combines with (4.11) to yield the Triple Limit Theorem (Theorem 4.8(b)):

$$\lim_{b\to\infty} \lim_{a\to-\infty} \lim_{N\to\infty} \gamma_n^N(a, b) = \gamma_n; \lim_b \lim_a \lim_N v_n^N(a, b) = v_n \qquad (n = 1, 2, \ldots).$$

In the course of our investigations it is convenient to introduce the concept of an *extended* s.v., i.e., to drop the requirement that $t$ be finite with probability one, while defining $x_\infty = \limsup_{n\to\infty} x_n$. More formally, if $\{y_n, \mathscr{F}_n\}_1^\infty$ is any stochastic sequence, we say that $t$ is an *extended* s.v. if $t$ is a random variable taking values in $\{1, 2, \ldots, +\infty\}$ such that $\{t = n\} \in \mathscr{F}_n$ $(n = 1, 2, \ldots)$. By $\{t = \infty\}$ we mean $\bigcap_{n=1}^{\infty} \{t > n\}$. We put $y_\infty = \limsup_{n\to\infty} y_n$ and agree that $y_t = y_\infty$ on $\{t = \infty\}$. Returning to the basic integrable stochastic sequence $\{x_n, \mathscr{F}_n\}_1^\infty$, let $\bar{C}$ denote the class of all extended s.v.'s such that $Ex_t^- < \infty$, $\bar{C}_n$ = all $t \in \bar{C}$ such that $t \geq n$, $\bar{\gamma}_n = \operatorname{ess\,sup}_{t\in\bar{C}_n} E(x_t \mid \mathscr{F}_n)$ $(n = 1, 2, \ldots)$. Note that according to our definitions $\bar{\gamma}_\infty = \limsup_{n\to\infty} \bar{\gamma}_n$. It is not true in general that $\bar{\gamma}_\infty = x_\infty$ (however, see Lemma 4.10).

The notion of an extended s.v. will prove a useful technical device in what follows, motivating methods of proof and unifying results within the usual framework. Section 4.6 (see also Section 5.2(c)) contains an example which suggests that such rules may be of interest in themselves. The following version of Lemma 3.3 is relevant to the study of extended s.v.'s.

**Lemma 4.8** Assume that $\{y_n, \mathscr{F}_n\}_1^\infty$ is an integrable stochastic sequence and $t$ an extended s.v. such that

$$E(y_{n+1} \mid \mathscr{F}_n) \geq y_n \quad \text{on} \quad \{t > n\} \qquad (n = 1, 2, \ldots).$$

Then $\{y_{\min(t,n)}, \mathscr{F}_n, 1 \leq n < \infty\}$ is a submartingale. If in addition $Ey_t$ exists and $(I\{t > n\}y_n^+)$ is uniformly integrable, then

$$E(y_t \mid \mathscr{F}_n) \geq y_n \quad \text{on} \quad \{t \geq n\} \quad (n = 1, 2, \ldots). \tag{4.12}$$

***Proof*** The first part follows as in the proof of Lemma 3.3. To prove (4.12) note that for arbitrary $A \in \mathscr{F}_n$, $N = n + 1, n + 2, \ldots$

$$\int_{A(t\geq n)} y_n \leq \int_{A(t\geq n)} y_{\min(t,N)} = \int_{A(n\leq t\leq N)} y_t + \int_{A(t>N)} y_N.$$

Letting $N \to \infty$ we have by the assumed uniform integrability of $(I\{t > n\}y_n^+)$ and Fatou's lemma (Lemma 1.2)

$$\int_{A(t \geq n)} y_n \leq \int_{A(n \leq t < \infty)} y_t + \int_{A(t=\infty)} y_\infty = \int_{A(t \geq n)} y_t,$$

from which (4.12) follows.

**Lemma 4.9** If $\{\beta_n, \mathscr{F}_n\}_1^\infty$ is a stochastic sequence such that for each $n \geq 1$

(a) $x_n \leq \beta_n \leq E(u \mid \mathscr{F}_n)$ for some non-negative random variable $u$ with finite expectation,

(b) $$\beta_n \leq \max(x_n, E(\beta_{n+1} \mid \mathscr{F}_n)),$$

and

(c) $$\beta_\infty = x_\infty,$$

then

$$\beta_n \leq \gamma_n \qquad (n = 1, 2, \ldots).$$

***Proof*** Let $n = 1, 2, \ldots$ be arbitrary. It suffices to show

$$\int_A \beta_n \leq \int_A \gamma_n \ (A \in \mathscr{F}_n).$$

Suppose on the contrary there exists an $\varepsilon > 0$, $A \in \mathscr{F}_n$, such that

$$\int_A \beta_n > \int_A \gamma_n + \varepsilon. \tag{4.13}$$

Define

$$t = \text{first } k \geq n \quad \text{such that } x_k \geq \beta_k - \varepsilon$$
$$= \infty \quad \text{if no such } k \text{ exists.}$$

From (a) and (b) we see that Lemma 4.8 applies to $\{\beta_n, \mathscr{F}_n\}_1^\infty$ and $t$, and hence by (c), we have for any $B \in \mathscr{F}_n$

$$-\infty < \int_B \beta_n \leq \int_{B(t<\infty)} \beta_t + \int_{B(t=\infty)} x_\infty. \tag{4.14}$$

Putting $B = \Omega$, we see that $P\{t = \infty, x_\infty = -\infty\} = 0$. But by (c) $P\{t = \infty, x_\infty > -\infty\} = 0$. Hence $P\{t < \infty\} = 1$ and $t \in C_n$; and from (4.13), (4.14) with $B = A$, and the definition of $t$ we have

$$\int_A \gamma_n < \int_A \beta_n - \varepsilon \leq \int_A x_t = \int_A E(x_t \mid \mathscr{F}_n),$$

contradicting the definition of $\gamma_n$.

**Lemma 4.10** If $E(\sup x_n^+) < \infty$ and $\{\beta_n, \mathscr{F}_n\}_1^\infty$ is a stochastic sequence such that for each $n = 1, 2, \ldots, x_n \leq \beta_n \leq E(\sup_{k \geq n} x_k \mid \mathscr{F}_n)$, then $\beta_\infty = x_\infty$.

***Proof*** Clearly $\beta_\infty \geq x_\infty$. Moreover, for each $m = 1, 2, \ldots, n$

$$\beta_n \leq E(\sup_{k \geq m} x_k \mid \mathscr{F}_n),$$

and hence by Lévy's theorem (Theorem 1.4)

$$\beta_\infty \leq \sup_{k \geq m} x_k \to x_\infty \quad \text{as} \quad m \to \infty.$$

For each $N = 1, 2, \ldots$, define

$$\text{(4.15)} \qquad \begin{aligned} \tilde{\gamma}_N^N &= E(\sup_{k \geq N} x_k \mid \mathscr{F}_N) \\ \tilde{\gamma}_n^N &= \max(x_n, E(\tilde{\gamma}_{n+1}^N \mid \mathscr{F}_n)) \qquad (n = N-1, \ldots, 1). \end{aligned}$$

It is easily seen by induction that for each $n$, $\tilde{\gamma}_n^n \geq \tilde{\gamma}_n^{n+1} \geq \cdots$ and hence we may define $\tilde{\gamma}_n = \lim_{N \to \infty} \tilde{\gamma}_n^N$.

**Lemma 4.11** If $E(\sup x_n^+) < \infty$, then for each $n = 1, 2, \ldots$

(a) $$\tilde{\gamma}_n = \bar{\gamma}_n = \gamma_n,$$

(b) $$\lim_{a \to -\infty} \gamma_n(a) = \gamma_n.$$

***Proof*** (a) It is easily seen that

$$\tilde{\gamma}_n \geq \bar{\gamma}_n \geq \gamma_n \qquad (n = 1, 2, \ldots).$$

In fact, as in the proof of Theorem 4.1 we have

$$\text{(4.16)} \qquad \bar{\gamma}_n \leq \max(x_n, E(\bar{\gamma}_{n+1} \mid \mathscr{F}_n)).$$

But $\bar{\gamma}_N \leq \tilde{\gamma}_N^N$ for all $N$, and by (4.15) and (4.16) we see that $\bar{\gamma}_n \leq \tilde{\gamma}_n^N$ $(N = n, n+1, \ldots, n = 1, 2, \ldots)$ and hence $\bar{\gamma}_n \leq \tilde{\gamma}_n$ $(n = 1, 2, \ldots)$. By (4.15) and the monotone convergence theorem for conditional expectations

$$\tilde{\gamma}_n = \max(x_n, E(\tilde{\gamma}_{n+1} \mid \mathscr{F}_n)) \qquad (n = 1, 2, \ldots),$$

and we complete the proof of (a) by appealing to Lemmas 4.9 and 4.10.

(b) $\gamma_n(a)$ is an increasing function of $a$ for each $n$; let $\gamma_n^* = \lim_{a \to -\infty} \gamma_n(a)$.

Then $\gamma_n^* \geq \gamma_n$ and by the monotone convergence theorem for conditional expectations

$$\gamma_n^* = \max(x_n, E(\gamma_{n+1}^* \mid \mathscr{F}_n)) \qquad (n = 1, 2, \ldots).$$

From Lemma 4.10 we see that $x_\infty(a) = \gamma_\infty(a)$ for all $a > -\infty$. But $\gamma_\infty^* \leq \gamma_\infty(a) = x_\infty(a) = \max(x_\infty, a) \downarrow x_\infty$ as $a \downarrow -\infty$, and the proof is completed by appealing to Lemma 4.9.

The above lemma says, among other things, that if we relax the definition of a s.v. in the prescribed manner, then $(\gamma_n)$ (and hence $(v_n)$) is not increased, provided $E(\sup x_n^+) < \infty$. That this result holds without such a restriction is a consequence of

**Lemma 4.12**

$$\lim_{b\to\infty} \bar{\gamma}_n(b) = \bar{\gamma}_n \qquad (n = 1, 2, \ldots).$$

***Proof*** Let $n = 1, 2, \ldots$ be arbitrary. $\bar{\gamma}_n(b)\uparrow$ as $b \uparrow \infty$; let $\gamma_n^* = \lim_{b\to\infty} \bar{\gamma}_n(b)$. Then $\gamma_n^* \leq \bar{\gamma}_n$. For each $b > 0$, $t \in \bar{C}_n$

$$E(x_t(b) \mid \mathscr{F}_n) \leq \bar{\gamma}_n(b) \leq \gamma_n^*.$$

But since $x_t^-(b) = x_t^-$ and $Ex_t^- < \infty$ for each $t \in \bar{C}_n$, we have by the monotone convergence theorem for conditional expectations

$E(x_t \mid \mathscr{F}_n) = \lim_{b\to\infty} E(x_t(b) \mid \mathscr{F}_n) \leq \gamma_n^*$ $(t \in C_n)$ and hence $\bar{\gamma}_n \leq \gamma_n^*$.

***Remark*** Lemma 4.12 makes it easy to complete the proof of Theorem 4.6.

We summarize the foregoing results as

**Theorem 4.7** For each $n = 1, 2, \ldots$

$$\gamma_n = \operatorname*{ess\,sup}_{t \in \bar{C}_n} E(x_t \mid \mathscr{F}_n),\ v_n = \sup_{\bar{C}_n} Ex_t,$$

and

**Theorem 4.8** For each $n = 1, 2, \ldots$

(a) $$\gamma_n = \lim_{b\to\infty} \lim_{N\to\infty} \tilde{\gamma}_n^N(b),\ v_n = \lim_b \lim_N E\tilde{\gamma}_n^N(b),$$

(b) Triple Limit Theorem:

$$\gamma_n = \lim_{b\to\infty} \lim_{a\to-\infty} \lim_{N\to\infty} \gamma_n^N(a, b),\ v_n = \lim_b \lim_a \lim_N v_n^N(a, b);$$

in particular, $V = \lim_b \lim_a \lim_N V^N(a, b)$.

In spite of its forbidding appearance, the complete generality of Theorem 4.8 makes it quite useful. Chapter 5 contains several applications.

The method of proof of Theorem 4.8 also yields the more general

**Theorem 4.9** Let $\{x_n(p), \mathscr{F}_n\}_1^\infty$ be a family of integrable stochastic sequences indexed by an extended real parameter $p$ assuming values in some closed (perhaps infinite) interval $[a, b]$. Let $p_0 \in [a, b]$ and assume that $x_n(p) \uparrow\downarrow x_n(p_0)$ as $p \uparrow\downarrow p_0$ $(n = 1, 2, \ldots)$.
(a) If there exists a $p^* > p_0$ such that

$$E(\sup_n x_n^+(p^*)) < \infty$$

and if

$$x_\infty(p) \downarrow x_\infty(p_0) \quad \text{as} \quad p \downarrow p_0,$$

then as $p \downarrow p_0$

$$\gamma_n(p) \downarrow \gamma_n(p_0),\ v_n(p) \downarrow v_n(p_0) \qquad (n = 1, 2, \ldots).$$

(b) If there exists a $p_* < p_0$ such that

$$\gamma_n(p) = \operatorname*{ess\,sup}_{t \in C_n(p_*)} E(x_t(p) \mid \mathscr{F}_n) \quad (p_* \le p \le p_0),$$

then

$$\gamma_n(p) \uparrow \gamma_n(p_0),\ v_n(p) \uparrow v_n(p_0) \quad \text{as} \quad p \uparrow p_0 \qquad (n = 1, 2, \ldots).$$

An analogous statement holds if $x_n(p)$ is monotonically decreasing in $p$ for each $n$.

An immediate application of Theorem 4.7 is the following generalization of Theorem 4.5. We extend this idea below to give necessary and sufficient conditions for $\sigma$ to be optimal in $\bar{C}$.

**Theorem 4.5′** If $E(\sup x_n^+) < \infty$, then $\sigma$ is optimal in $\bar{C}$. If in addition $x_n \to -\infty$, then $\sigma \in C$.

***Proof*** The proof of Theorem 4.5 shows that $Ex_\sigma \ge V$. Since by Theorem 4.7 $Ex_t \le V$ $(t \in \bar{C})$ we see that $\sigma$ is optimal in $\bar{C}$. If $x_n \to -\infty$ it is clear that $\sigma$ cannot assume the value $+\infty$ (and collect a reward of $-\infty$) with positive probability.

**Theorem 4.10** Let $\sigma(n)$ denote $\min(\sigma, n)$ $(n \ge 1)$. If $V < \infty$ the following statements are equivalent.
(a) $\sigma$ is optimal in $\bar{C}$.
(b) $(\gamma_{\sigma(n)})$ is uniformly integrable.

(c) $(\gamma_{\sigma(n)}^{+})$ is uniformly integrable.
(d) $\int_{(\sigma<\infty)} x_\sigma^+ < \infty$ and $(I\{\sigma > n\}\gamma_n^+) = (I\{\sigma > n\}E^+(\gamma_{n+1} \mid \mathscr{F}_n))$ is uniformly integrable.

***Proof*** If $\sigma$ is optimal in $\overline{C}$, then (4.7) holds by an argument identical to that given in the proof of Theorem 4.2, and hence

$$E(x_\sigma \mid \mathscr{F}_n) = \gamma_{\sigma(n)} \qquad (n \geq 1).$$

By Lemma 2.2, (b) holds, and thus by the definition of uniform integrability so does (c). The equivalence of (c) and (d) is easily verified. We now show that (c) implies (a). By Lemma 4.8

$$\{\gamma_{\sigma(n)}, \mathscr{F}_n, 1 \leq n < \infty\}$$

is a martingale and hence $\{\gamma_{\sigma(n)}^+, \mathscr{F}_n, 1 \leq n < \infty\}$ is a submartingale. Let $\tilde{x}_n = \gamma_{\sigma(n)}^+$ $(n \geq 1)$. Then for any s.v. $t$ and $N = 1, 2, \ldots$

$$\int_{(t \leq N)} \tilde{x}_t = \sum_1^N \int_{(t=k)} \tilde{x}_k \leq \sum_1^N \int_{(t=k)} \tilde{x}_N \leq E\tilde{x}_N,$$

so letting $N \to \infty$ we have

$$E\tilde{x}_t \leq \sup_N E\tilde{x}_N < \infty;$$

and thus $\tilde{V} < \infty$. By Theorem 4.7

$$E\gamma_\sigma^+ = E\tilde{x}_\infty < \infty \quad \text{and hence} \quad Ex_\sigma^+ < \infty.$$

Now for any $n = 1, 2, \ldots, A \in \mathscr{F}_n$

$$\int_A \gamma_{\sigma(n)}^+ \leq \int_A \gamma_{\sigma(m)}^+ \qquad (m = n, n+1, \ldots),$$

so by Fatou's lemma (Lemma 1.2)

$$\text{(4.17)} \qquad \int_A \gamma_{\sigma(n)}^+ \leq \limsup_{m\to\infty} \int_A \gamma_{\sigma(m)}^+ \leq \int_A \gamma_\sigma^+.$$

By Lemmas 4.1–4.5 there exists a sequence of s.v.'s $t_1 \leq t_2 \leq \cdots \leq \sigma$ such that $Ex_{t_i} \uparrow V$. By Lemma 4.6 $\lim_{i\to\infty} t_i = \sigma$. For each $c > 0$ we have by (4.17)

$$\begin{aligned} cP\{\gamma_{t_i} > c\} &\leq \int_{\{\gamma_{t_i} > c\}} \gamma_{t_i} = \sum_{n=1}^{\infty} \int_{\{t_i=n, \gamma_n > c\}} \gamma_n \\ &= \sum_{n=1}^{\infty} \int_{\{t_i=n,\gamma_n>c\}} \gamma_{\sigma(n)}^+ \leq \sum_{n=1}^{\infty} \int_{\{t_i=n,\gamma_n>c\}} \gamma_\sigma^+ \\ &= \int_{\{\gamma_{t_i}>c\}} \gamma_\sigma^+ \leq E\gamma_\sigma^+ < \infty, \end{aligned}$$

and it follows (see the proof of Lemma 2.2) that $(\gamma_{t_i}^+)$ and a fortiori $(x_{t_i}^+)$ is uniformly integrable. Thus by Fatou's lemma

$$Ex_\sigma \geq \limsup_{i\to\infty} Ex_{t_i} = V.$$

Since by Theorem 4.7 $Ex_t \leq V$ $(t \in \bar{C})$, it follows that $\sigma$ is optimal in $\bar{C}$.

***Remark*** The preceding proof that (c) implies (a) may be shortened by appropriate reference to Theorem 2.1. A somewhat different proof, also using the martingale convergence theorem, is as follows. By Theorem 2.1 $\lim_{n\to\infty} \gamma_{\sigma(n)}$ exists (i.e., $\gamma_n \to \gamma_\infty$ on $\{\sigma = \infty\}$) and

$$E\gamma_\sigma \geq \lim_{n\to\infty} E\gamma_{\sigma(n)} = E\gamma_1 = V. \tag{4.18}$$

Hence by Lemma 4.13 below $x_\infty = \gamma_\infty$ on $\{\sigma = \infty\}$, and since by the definition of $\sigma$, $x_\sigma = \gamma_\sigma$ on $\{\sigma < \infty\}$ we have from (4.18)

$$Ex_\sigma \geq V.$$

**Lemma 4.13** If $V < \infty$, then for each $\varepsilon > 0$

$$P\{x_n \geq \gamma_n - \varepsilon \text{ i.o.}\} = 1$$

where i.o. abbreviates infinitely often. (See Problem 7.)

***Proof*** Denote by $B = B(m, \varepsilon)$ the event that $x_n \leq \gamma_n - \varepsilon$ for all $n \geq m$, and suppose by way of contradiction that for some $\varepsilon > 0$, $m = 1, 2, \ldots, P(B) > 0$. For all $t \in C_m$

$$\int_B x_t \leq \int_B \gamma_t - \varepsilon P(B).$$

But $\int_{\bar{B}} x_t \leq \int_{\bar{B}} \gamma_t$ and thus by Theorem 4.6 for all $t \in C_m$

$$Ex_t \leq E\gamma_t - \varepsilon P(B) \leq E\gamma_m - \varepsilon P(B),$$

contradicting Theorem 4.1.

## 6. Examples and Counter-examples

(a) Our first example shows that (4.11) is not true in general. For future reference we construct this example in complete detail. Let $a_1, a_2, \ldots, b_1, b_2, \ldots$ be two increasing sequences of non-negative real numbers and $\Omega$ the space of sequences

$$\omega_j = (a_1, \ldots, a_j, -b_{j+1}, -b_{j+2}, \ldots) \qquad (j = 1, 2, \ldots).$$

Let $\mathscr{F}$ be all subsets of $\Omega$ and $P(\omega_j) = 1/j - 1/(j+1)$ $(j = 1, 2, \ldots)$. For each $n = 2, 3, \ldots$ let $x_n(\omega_j)$ be the $n$th coordinate of $\omega_j$ and let $\mathscr{F}_n = \mathscr{B}(x_2, \ldots, x_n)$. It is easily seen that for each $n$

$$P(x_n = a_n) = \frac{1}{n} = 1 - P(x_n = -b_n),$$

$$E(x_{n+1} \mid x_n = a_n) = a_{n+1}\frac{n}{n+1} - b_{n+1}\left(1 - \frac{n}{n+1}\right),$$

$$Ex_n = \frac{a_n}{n} - b_n\left(1 - \frac{1}{n}\right).$$

Let $t$ = first $k \geq 2$ such that $x_k = -b_k$ and $t_n = \min(t, n)$ $(n = 2, 3, \ldots)$. Obviously the rules $(t_n)$ are the only ones which need be considered in searching for an optimal rule or in computing the value of $\{x_n, \mathscr{F}_n\}_2^\infty$. Straightforward calculations give

$$Ex_{t_n}^+ = a_n/n, \quad Ex_{t_n}^- = \sum_{j=2}^{n} b_j/j(j-1).$$

Now put $a_j = j(j-1) = b_j$ $(j = 1, 2, \ldots)$. Then $Ex_{t_n}^+ \to +\infty$ and in the notation of the preceding section $V(a) = \infty$, all $a \neq -\infty$. But $Ex_{t_n} = n - 1 - \sum_2^n 1 = 0$, $V = 0$, and it is not true that $\lim_{a \to -\infty} V(a) = V$.

(*b*) Our next example shows the necessity of some condition like $E(\sup x_n^+) < \infty$ in Theorem 4.5 (see also Theorem 4.10). With the same stochastic sequence as in (a) let $a_n = n^2 - 1$, $b_n = n(n-1)$. Then $x_n \to -\infty$, $Ex_n \to -\infty$, but $Ex_{t_n} = 1 - 1/n$ and no optimal rule exists.

(c) One-armed bandit with discounting. Our last example is a Bayesian decision problem in which there is no optimal rule in $C$ even though $\sigma$ is optimal in $\bar{C}$.

Suppose that conditionally on $p$ $(0 < p < 1)$ $y_1, y_2, \ldots$ are i.i.d. with $P\{y_1 = 1\} = p = 1 - P\{y_1 = -1\}$. For some $0 < \alpha < 1$ and each $n = 0, 1, 2, \ldots$ let $x_n = \sum_{k=1}^{n} \alpha^{k-1} y_k$, $\mathscr{F}_n = \mathscr{B}(y_1, \ldots, y_n)$, and assume that there exists a known prior distribution of $p$. By Theorem 4.5′ $\sigma$ is optimal in $\bar{C}$. Suppose for simplicity that the prior distribution of $p$ is a member of the beta family with parameters $(r, q)$. Let $S_n = y_1^+ + \cdots + y_n^+$. It is well-known (and easily verified) that the posterior distribution of $p$ after observing $y_1, \ldots, y_n$ is beta with parameters $(r + S_n, q + n - S_n)$. Letting $V(r, q)$ denote the value of the sequence $\{x_n, \mathscr{F}_n\}_0^\infty$ as a function of $(r, q)$, it is plausible and follows rigorously from Theorem 5.2 that

$$\sigma = \text{first } n \geq 0 \quad \text{such that } V(r + S_n, q + n - S_n) \leq 0$$

$$(= \infty \quad \text{if no such } n \text{ exists}).$$

Now $E(x_1 \mid p) = 2p - 1$ and hence $V(r, q) \geq (r - q)/(r + q)$, where the inequality follows from considering the rule $t = 1$ and the fact that the expectation of a beta random variable with parameters $(r, q)$ is $r/(r + q)$. Thus if $r > q$, for $\sigma$ to be infinite it suffices that $S_n > n/2$ for all $n \geq 1$. But this occurs with positive probability when $p > \frac{1}{2}$, and hence if $r > q$, $P\{\sigma = \infty\} > 0$.

## 7. Optimal Stopping for $s_n/n$

Let $y_1, y_2, \ldots$ be independent random variables with mean 0 and variance 1. Put $s_n = \sum_1^n y_k$, $\mathscr{F}_n = \mathscr{B}(y_1, \ldots, y_n)$ $(n \geq 1)$. In this section we shall discuss the question of the existence of an optimal stopping rule for $\{s_n/n, \mathscr{F}_n\}_1^\infty$. We show that $\sigma$ is optimal in $\overline{C}$, and if condition (4.19) below is satisfied, then $\sigma \in C$. (This condition is always satisfied in the i.i.d. case—see Remark (i) at the end of this section.)

Let $s_0 = 0$, $\mathscr{F}_0 = [\phi, \Omega]$, and $s_k^{(n)} = s_k - s_n$ $(n \geq 0, k \geq n)$.

**Theorem 4.11** Let $(y_n)$, $(s_n)$, $(\mathscr{F}_n)$ be as above. Then $V < \infty$ and $\sigma$ is optimal in $\overline{C}$ for $\left\{\frac{s_n}{n}, \mathscr{F}_n\right\}_1^\infty$. If

$$P\{\limsup_n n^{-1/2} s_n > 0\} > 0, \tag{4.19}$$

then $\sigma \in C$.

The proof of Theorem 4.11 will be preceded by two lemmas.

**Lemma 4.14**

$$E(\sup |s_n/n|) < \infty.$$

Moreover, for each $1 \leq p < 2$, $n = 1, 2 \ldots$

$$E(\sup_{k>n} |s_k^{(n)}/k|^p) \leq \frac{2}{2 - p} n^{-p/2}. \tag{4.20}$$

***Proof*** Let $n = 1, 2, \ldots$ be fixed. By Section 2.4(b) applied to the submartingale $\{s_k^{(n)2}, \mathscr{F}_k\}_{n+1}^{\infty}$, we have

$$E(\sup_{k>n} |s_k^{(n)}/k|^p) \leq p \int_0^{\infty} u^{p-1} P\{|s_k^{(n)}| > ku \quad \text{for some} \quad k > n\}\, du$$

$$\leq n^{-(p/2)} + p \int_{1/\sqrt{n}}^{\infty} \left(\sum_{n+1}^{\infty} k^{-2}\right) u^{p-3}\, du$$

$$\leq \frac{2}{2-p} n^{-(p/2)}.$$

**Lemma 4.15** For each $t \in \overline{C}$, $\varepsilon > 0$, $1 < p < 2$, $n = 1, 2, \ldots$

$$E(s_t/t \mid \mathscr{F}_n) < s_n/n$$

on

$$A_n(t) = \left\{ s_n \geq \varepsilon n^{1/2}, P\{t = \infty \mid \mathscr{F}_n\} > \tfrac{1}{2}, \right.$$
$$\left. P\{n \leq t < \infty \mid \mathscr{F}_n\} < \left(\frac{2-p}{2}\right)^{1/(p-1)} \left(\frac{\varepsilon}{2}\right)^{p/(p-1)} \right\}$$

***Proof*** By the strong law of large numbers (Problem 2.14) $s_t/t = 0$ on $\{t = \infty\}$. Hence by Lemma 4.14 and Hölder's inequality, we have on $A_n(t)$

$$E(s_t/t \mid \mathscr{F}_n) = s_n E(t^{-1} \mid \mathscr{F}_n) + E\left(\frac{s_t^{(n)}}{t} I_{\{n \leq t < \infty\}} \mid \mathscr{F}_n\right)$$

$$\leq \frac{s_n}{n}(1 - P\{t = \infty \mid \mathscr{F}_n\})$$

$$+ \left[E\left(\sup_{k>n} \left|\frac{s_k^{(n)}}{k}\right|^p\right)\right]^{1/p}$$

$$\times [P\{n \leq t < \infty \mid \mathscr{F}_n\}]^{(p-1)/p}$$

$$< \frac{s_n}{n} - \frac{\varepsilon n^{1/2}}{n} \cdot \frac{1}{2} + \left(\frac{2}{2-p}\right)^{1/p} n^{-1/2} \left(\frac{2-p}{2}\right)^{1/p} \frac{\varepsilon}{2}$$

$$\leq s_n/n.$$

***Proof of Theorem 4.11*** By Lemma 4.14 and Theorem 4.5′, $V < \infty$ and $\sigma$ is optimal in $\overline{C}$. It remains to show that $P\{\sigma < \infty\} = 1$.

Suppose the contrary. Let

$$t = \text{first } n \geq 1 \quad \text{such that} \quad A_n(\sigma) \text{ occurs}$$
$$= \infty \quad \text{if} \quad \bar{A}_n(\sigma) \text{ occurs for all } n,$$

where $A_n(\sigma)$ is defined as in Lemma 4.15. It is easy to see from Lévy's theorem (Theorem 1.4) that $P\{n \leq \sigma < \infty \mid \mathscr{F}_n\} \to 0$ and on $\{\sigma = \infty\}$, $P\{\sigma = \infty \mid \mathscr{F}_n\} \to 1$ as $n \to \infty$. Also by (4.19) and the Kolmogorov $0-1$ law there exists an $\varepsilon > 0$ such that $P\{\limsup_n n^{-1/2} s_n > \varepsilon\} = 1$. Hence $t$ is finite on $\{\sigma = \infty\}$. Let $t' = \min(t, \sigma)$. Then by Lemma 4.15

$$\int_{\{t<\sigma\}} s_\sigma/\sigma = \sum_1^\infty \int_{\{t=n<\sigma\}} E(s_\sigma/\sigma \mid \mathscr{F}_n) < \sum_1^\infty \int_{\{t=n<\sigma\}} s_n/n$$
$$= \int_{\{t<\sigma\}} s_t/t.$$

Thus $E(s_{t'}/t') > E(s_\sigma/\sigma)$, a contradiction.

Modifications of the preceding method allow one to treat other reward sequences of the form $h_n(s_n)$. We indicate below a slightly different approach, which may in some cases have advantages.

To illustrate the second approach, we shall apply it to the stochastic sequence $\left\{\dfrac{s_n^2}{n^\alpha}, \mathscr{F}_n\right\}_1^\infty$, where $\alpha > 1$. Since $s_n^2 \geq 0$ for all $n$, to show that $\sigma$ is optimal in $\bar{C}$ it suffices by Theorem 4.10 to show that

$$\lim_{n\to\infty} E\gamma_n = 0. \tag{4.22}$$

The following lemma provides suitable upper bounds for the sequence $(\gamma_n)$.

**Lemma 4.16** For every $n = 0, 1, 2, \ldots$ and for every extended s.v. $t$

$$E\left(\frac{s_t^{(n)2}}{t^\alpha} \,\middle|\, \mathscr{F}_n\right) \leq \sum_{n+1}^\infty k^{-\alpha} < \infty \quad \text{on} \quad \{t > n\}. \tag{4.23}$$

In particular for every $n = 1, 2, \ldots$

$$E\left(\frac{s_t^{(n)2}}{t^\alpha} \,\middle|\, \mathscr{F}_n\right) \leq (\alpha - 1)^{-1} n^{1-\alpha} \quad \text{on} \quad \{t > n\}. \tag{4.24}$$

***Proof*** By Theorem 4.7 we may assume that $t \in C$. Since by Lemma 2.1 $\{s_k^{(n)2}, \mathscr{F}_k\}_{n+1}^{\infty}$ is a submartingale, for each fixed $n = 0, 1, \ldots, A \in \mathscr{F}_n$, putting $\tilde{A} = A(t > n)$ we have

$$\text{(4.25)} \quad \int_{\tilde{A}} \frac{s_t^{(n)2}}{t^\alpha} \leq \sum_{n+1}^{\infty} k^{-\alpha} \left[ \int_{\tilde{A}(t \leq k)} s_k^{(n)2} - \int_{\tilde{A}(t \leq k-1)} s_{k-1}^{(n)2} \right]$$

$$= \sum_{n+1}^{\infty} k^{-\alpha} c_k,$$

say, where for each $N = n + 1, n + 2, \ldots$

$$\text{(4.26)} \quad \sum_{n+1}^{N} c_k \leq P(\tilde{A}) E s_N^{(n)2} = P(\tilde{A})(N - n).$$

Since $k^{-\alpha}$ is strictly decreasing the last term of (4.25) is a maximum subject to (4.26) if $c_{n+1} = c_{n+2} = \cdots = P(\bar{A})$. This proves (4.23). To prove (4.24) we bound the right-hand side of (4.23) by $\int_0^\infty (n + x)^{-\alpha}\, dx$.

We are now in a position to show that $\sigma$ is optimal in $\bar{C}$ for $\left\{\frac{s_n^2}{n^\alpha}, \mathscr{F}_n\right\}_1^\infty$. By (4.23) with $n = 0$ we see that $V < \infty$. To verify (4.22), observe that by Lemma 4.16 for each $t \in C_{n+1}$

$$E\left(\frac{s_t^2}{t^\alpha} \,\middle|\, \mathscr{F}_n\right) \leq 2\left(\frac{s_n^2}{n^\alpha} + (\alpha - 1)^{-1} n^{1-\alpha}\right) \quad (n = 1, 2, \ldots).$$

Hence

$$\gamma_n = \max\left(\frac{s_n^2}{n^\alpha}, \quad E(\gamma_{n+1} \mid \mathscr{F}_n)\right)$$

$$\leq 3\frac{s_n^2}{n^\alpha} + 2(\alpha - 1)^{-1} n^{1-\alpha},$$

from which (4.22) follows.

If for every $K > 0$

$$\text{(4.27)} \quad P\{\limsup_n n^{-(1/2)}|s_n| > K\} = 1,$$

then we can show that $P\{\sigma < \infty\} = 1$ by an argument similar to that of Theorem 4.11.

***Remarks*** (i) Conditions (4.19) and (4.27) hold if $y_1, y_2, \ldots$ are i.i.d. or more generally if, for example,

$$\lim_{n \to \infty} P\left\{\frac{s_n}{\sqrt{n}} \leq x\right\} = \frac{1}{\sqrt{2\pi}} \int_{-\infty}^{x} e^{-(1/2)y^2}\, dy \; (-\infty < x < \infty).$$

Then for any $K, p \equiv P\{\limsup_{n\to\infty} \frac{s_n}{\sqrt{n}} > K\} \geq \limsup_{n\to\infty} P\left\{\frac{s_n}{\sqrt{n}} > K\right\} >$ 0, and hence by the Kolmogorov 0 − 1 law (Section 2.2(b)), $p = 1$.

(ii) Results about $\left|\frac{s_n}{n}\right| \left(\frac{s_n}{n}\right)$ may easily be inferred from the corresponding results about $\left(\frac{s_n}{n}\right)^2 \left(\left(\frac{s_n^+}{n}\right)^2\right)$. In fact if $\{x_n, \mathscr{F}_n\}_1^\infty$ is any integrable stochastic sequence for which $V < \infty$ and $\sigma$ is optimal in $\bar{C}(C)$, and if $g$ is any non-decreasing concave function, then $V_g < \infty$ and $\sigma_g$ is optimal in $\bar{C}(C)$ for $\{g(x_n), \mathscr{F}_n\}_1^\infty$.

## 8. The Conditions $V < \infty$ and $E[\sup x_n^+] < \infty$

Theorem 4.5′ states that if

$$E[\sup x_n^+] < \infty, \tag{4.29}$$

then $\sigma$ is optimal in $\bar{C}$. However, our method of proof of Theorem 3.1 and the application of this theorem in Section 3.6(a), as well as the techniques of the preceding section, show that in some cases we can verify that

$$V < \infty \tag{4.30}$$

and that $\sigma$ is optimal in $\bar{C}$ without appealing to Theorem 4.5′. Obviously (4.29) implies (4.30). In this section we indicate the extent to which the converse is true.

(To see that (4.30) does not imply (4.29) in general, let $\{x_n, \mathscr{F}_n\}_1^\infty$ be a non-negative martingale which is not uniformly integrable. (See Sections 2.2(d) or 4.6(a) with $a_n = n, b_n = 0$ for examples of such martingales.) By Theorem 2.2 $V = Ex_1 < \infty$, but (4.29) does not hold.)

**Theorem 4.12** If $Ex_\infty^- < \infty$ and

$$\limsup_{a\to\infty} aP\{\sup_n x_n > a\} = +\infty, \tag{4.31}$$

then $V = +\infty$.

***Proof*** Let $a_m \to \infty$ be such that

$$a_m P\{\sup_n x_n > a_m\} \to \infty \qquad (m \to \infty).$$

For each $m = 1, 2, \ldots$ let

$$t(m) = \text{first } n \geq 1 \quad \text{such that } x_n > a_m$$
$$= \infty \quad \text{if no such } n \text{ exists.}$$

Then

$$V \geq Ex_{t(m)} \geq a_m P\{\sup_n x_n > a_m\} - Ex_\infty^- \to \infty \quad \text{as} \quad m \to \infty.$$

**Corollary** If $Ex_\infty^- < \infty$ and for some $0 < \alpha < 1$

$$E(\sup_n x_n^+)^\alpha = \infty,$$

then $V = \infty$.

***Proof*** If to the contrary $V < \infty$, then by Theorem 4.12 there exists a constant $c < \infty$ such that $\sup_a aP\{\sup_n x_n > a\} \leq c$. Then for any $0 < \alpha < 1$

$$E(\sup_n x_n^+)^\alpha = \alpha \int_0^\infty a^{\alpha-1} P\{\sup_n x_n > a\}\, da$$
$$\leq 1 + \alpha c \int_1^\infty a^{-2+\alpha}\, da < \infty,$$

contradicting the hypothesis of the corollary.

Theorem 4.12 and its corollary suggest that if $Ex_\infty^- < \infty$, then condition (4.30) is not appreciably weaker than (4.29). Theorem 4.14 shows that for the problems of $\frac{s_n}{n}$ and $\frac{y_n}{n}$ when $y_1, y_2, \ldots$ are i.i.d. (see Sections 4.7 and 5.8) the conditions (4.29) and (4.30) are equivalent. The following theorem, on the other hand, shows that for the stochastic sequences of Theorem 3.1 and Section 3.6(a) the statements (4.29) and (4.30) are appreciably different. In particular, in conjunction with Theorem 4.12 and its corollary, it provides another example of the phenomenon of Section 4.6(a), in which $\sup_t Ex_t^+ = \infty$ but $V < \infty$.

**Lemma 4.17** Let $y_1, y_2, \ldots$ be independent random variables with $Ey_n = 0$, $\sigma_n^2 = Ey_n^2 < \infty$, and put $s_n = \sum_1^n y_k$ $(n = 1, 2, \ldots)$. Then for any $a > 0$ and $n = 1, 2, \ldots$

$$P\{\max_{1 \leq k \leq n} s_k > a\} \leq 2P\left\{s_n > a - \left(2 \sum_1^n \sigma_k^2\right)^{1/2}\right\}.$$

***Proof*** Let $t$ denote the first $k \geq 1$ (if any) for which $s_k > a$, and put $b_n = \left(2 \sum_1^n \sigma_k^2\right)^{1/2}$. Then

$$\begin{aligned} P\{t \leq n, s_n \leq a - b_n\} &= \sum_{k=1}^{n} P\{t = k, s_n \leq a - b_n\} \\ &\leq \sum_{k=1}^{n} P\{t = k, s_n - s_k \leq -b_n\} \\ &\leq \sum_{k=1}^{n} P\{t = k\} P\{|s_n - s_k| \geq b_n\} \leq \tfrac{1}{2} P\{t \leq n\}, \end{aligned}$$

where the last inequality is a consequence of Chebyshev's inequality and the definition of $b_n$. Hence

$$\begin{aligned} P\{t \leq n\} &= P\{t \leq n, s_n > a - b_n\} + P\{t \leq n, s_n \leq a - b_n\} \\ &\leq P\{s_n > a - b_n\} + \tfrac{1}{2} P\{t \leq n\}, \end{aligned}$$

from which the lemma follows.

**Theorem 4.13** Let $y, y_1, y_2, \ldots$ be independent and identically distributed random variables with $Ey = 0$. Let $s_n = \sum_1^n y_k$ $(n \geq 1)$ and $\alpha > 0$. If

$$E[(y^+)^{1+\alpha}] < \infty, \tag{4.32}$$

then for every $a > 0$

$$E[\sup_n (s_n - na)^+]^\alpha < \infty \tag{4.33}$$

and

$$E[\sup_n (y_n - na)^+]^\alpha < \infty. \tag{4.34}$$

Conversely if either (4.33) or (4.34) holds for some $a > 0$, then (4.32) is satisfied.

***Proof*** We shall give the proof for the special case $\alpha = 1$. The general case is similar in principle, although the details are somewhat more complicated.

Assume that (4.32) holds. By passing to $y_n(c) = \max(y_n, -c)$, where $c$ is so large that $Ey(c) < a/2$, and noting that

$$s_n - na \leq \sum_1^n [y_k(c) - Ey(c)] - na/2 \qquad (n \geq 1),$$

we see that with no loss of generality we may assume that the sequence $y_1, y_2, \ldots$ is bounded from below. Define

$$y_n' = y_n I_{\{y_n \le n\}} \qquad y_n'' = y_n - y_n'$$
$$s_n' = \sum_1^n y_k' \qquad (n \ge 1).$$

To prove (4.33) (with $\alpha = 1$) it suffices to show

$$E\left(\sum_1^\infty y_n''\right) < \infty \tag{4.35}$$

and

$$E(\sup\,[s_n' - na]^+) < \infty. \tag{4.36}$$

By (4.32) we have

$$E\left(\sum_1^\infty y_n''\right) = \sum_{n=1}^\infty \sum_{k=n}^\infty \int_{\{k<y\le k+1\}} y = \sum_{k=1}^\infty k \int_{\{k<y\le k+1\}} y$$
$$\le E(y^+)^2 < \infty.$$

To complete the proof of (4.33) it remains to prove (4.36), and hence it suffices to show that for some $k_0 = 1, 2, \ldots$

$$\sum_{k=k_0}^\infty E\{\sup_{n_k \le n < n_{k+1}} (s_n' - na)^+\} < \infty,$$

where we have let $n_k$ denote the largest integer $\le \exp(k)$ $(k = 0, 1, \ldots)$. By Lemma 4.17, for all $k = 0, 1, \ldots,$ and $u > 0$

$$\begin{aligned} P\{\sup_{n_k \le n < n_{k+1}} (s_n' - na)^+ > u\} &\le P\left(\bigcup_{n=1}^{n_{k+1}} \{s_n' - Es_n' > u + n_k a\}\right) \\ &\le 2P\left\{s_{n_{k+1}}' - Es_{n_{k+1}}' \right. \\ &\qquad \left. > u + n_k a - \left(2\sum_1^{n_{k+1}} E(y_n')^2\right)^{1/2}\right\}. \end{aligned} \tag{4.37}$$

Since $\sum_1^n E(y_i')^2 \le (\text{const})\, n$ $(n \ge 1)$, we have by (4.37) for all $k_0$ sufficiently large

$$\begin{aligned} \sum_{k=k_0}^\infty E\{\sup_{n_k \le n < n_{k+1}} (s_n' - na)^+\} &\le \sum_{k_0}^\infty \int_0^\infty P\{\sup_{n_k \le n < n_{k+1}} (s_n' - na) > u\}\, du \\ &\le 2\sum_{k=k_0}^\infty \int_0^\infty P\{s_{n_{k+1}}' - Es_{n_{k+1}}' > u + a_1 n_{k+1}\}\, du, \end{aligned} \tag{4.38}$$

where we have set $a_1 = a/2e$. Now by the Markov and $c_r$ inequalities

$$
\begin{aligned}
P\{s'_{n_k} - Es'_{n_k} > u + a_1 n_k\} & \\
&\leq (u + a_1 n_k)^{-4} E[s'_{n_k} - Es'_{n_k}]^4 \\
&\leq (u + a_1 n_k)^{-4} \left[ \sum_1^{n_k} E(y'_j - Ey'_j)^4 + 0(n_k^2) \right] \\
&\leq 16(u + a_1 n_k)^{-4} \left[ \sum_1^{n_k} E(y'_j)^4 + 0(n_k^2) \right].
\end{aligned}
$$

Hence by (4.38)

$$
\begin{aligned}
\sum_{k=k_0}^{\infty} E\{ \sup_{n_k \leq n < n_{k+1}} (s'_n - na)^+ \} & \\
&\leq \text{const} \left[ \sum_{k=1}^{\infty} \sum_{j=1}^{n_k} n_k^{-3} E(y'_j)^4 + \sum_{k=1}^{\infty} n_k^{-3} 0(n_k^2) \right] \\
&\leq \text{const} \sum_{j=1}^{\infty} \sum_{k=[\log j]}^{\infty} n_k^{-3} E(y'_j)^4 \\
&\leq \text{const} \sum_{i=1}^{\infty} j^{-3} E(y'_j)^4 \leq \text{const} \sum_{j=1}^{\infty} \sum_{k=1}^{j} j^{-3} \int_{\{k-1<|y|\leq k\}} y^4 \\
&\leq \text{const} \sum_{k=1}^{\infty} k^{-2} \int_{\{k-1<|y|\leq k\}} y^4 \leq \text{const}\, Ey^2 < \infty.
\end{aligned}
$$

This completes the proof of (4.33) under the hypothesis (4.32).

Now let $a > 0$ and for any $k = 1, 2, \ldots$ let

$$A_k = \{y_k > 2(u + ak)\}, B_k = \{|s_{k-1}| < u + ak\}.$$

Then

$$
\begin{aligned}
P\left( \bigcup_1^{\infty} A_k \right) &\leq \sum_1^{\infty} P(A_k) = \sum_{k=1}^{\infty} \sum_{j=k}^{\infty} P\{aj < y/2 - u \leq a(j+1)\} \\
&= \sum_{j=1}^{\infty} jP\{aj < y/2 - u \leq a(j+1)\} \\
&\leq 1/aE(y/2 - u)^+ \to 0 \qquad (u \to \infty).
\end{aligned}
$$

Hence there exists a number $u_0$ such that for all $u \geq u_0$, $P\left( \bigcup_{k=1}^{\infty} A_k \right) < \frac{1}{3}$.

By the weak law of large numbers we may also assume that

$P(B_k) > \frac{2}{3}$ for all $u \geq u_0$ and $k = 1, 2, \ldots$. Hence for all $u \geq u_0$ and $n = 1, 2, \ldots$

$$\begin{aligned} P\{\max_{1 \le k \le n} (s_k - ak) > u\} &\geq P\left(\bigcup_1^\infty A_k B_k\right) \\ &\geq \sum_{k=1}^n P(\bar{A}_1 \cdots \bar{A}_{k-1} A_k B_k) \\ &\geq \sum_{k=1}^n \left[P(A_k B_k) - P\left(A_k \cap \bigcup_{i=1}^{k-1} A_i\right)\right] \\ &\geq \sum_{k=1}^n P(A_k)\left[P(B_k) - P\left(\bigcup_{i=1}^{k-1} A_i\right)\right] \\ &\geq \tfrac{1}{3} P\left(\bigcup_{k=1}^n A_k\right). \end{aligned}$$

Letting $n \to \infty$, we have for all $u$ sufficiently large

$$P\{\sup_n (s_n - na)^+ > u\} \geq \tfrac{1}{3} P\{\sup_n (y_n - 2na)^+ > 2u\},$$

and hence if (4.33) holds for some $a > 0$, then (4.34) holds for $2a$. To complete the proof of the theorem it suffices to show that if for some $a > 0$ (4.34) holds, then (4.32) is satisfied. In the case $\alpha = 1$, (4.34) implies that

$$\sum_{k=1}^\infty P\{\sup_n (y_n - na) > k\} < \infty,$$

or equivalently that

$$\prod_{k=1}^\infty \prod_{n=1}^\infty F(k + na) > 0,$$

where we have let $F$ denote the distribution function of $y$ and have assumed as we may by a change of scale that $F(1) > 0$. Hence

$$\sum_{k=1}^\infty \sum_{n=1}^\infty \log F(k + na) > -\infty,$$

and since $\log F(x) \sim -(1 - F(x))$ $(x \to \infty)$, we have

$$\begin{aligned} &\int_0^\infty \int_0^\infty (1 - F(x + ay))\, dx\, dy \\ &\qquad \leq \text{const} + \sum_1^\infty \sum_1^\infty (1 - F(k + na)) < \infty. \end{aligned}$$

Let $u = x + ay$. Then $\int_0^\infty \int_{ay}^\infty (1 - F(u))\, du\, dy < \infty$, or equivalently by Fubini's theorem

$$\int_0^\infty u(1 - F(u))\, du < \infty,$$

which in turn (in the case $\alpha = 1$) is equivalent to (4.32).

**Theorem 4.14** Let $y_1, y_2, \ldots$ be i.i.d. with $Ey_i = \mu$ for some $-\infty < \mu < \infty$, and put $s_n = \sum_1^n y_k$ $(n = 1, 2, \ldots)$. The following statements are equivalent:

(a) $E[y_1^+ \log^+ y_1^+] < \infty$,

(b) $E[\sup_n s_n/n] < \infty$,

(c) $\sup_{t \in C} E(s_t/t) < \infty$,

(d) $E(\sup y_n/n) < \infty$,

(e) $\sup_{t \in C} E(y_t/t) < \infty$.

***Proof*** It is obvious that (b) ⇒ (c) and (d) ⇒ (e). Hence it suffices to show (a) ⇒ (b), (a) ⇒ (d), (c) ⇒ (a), and (e) ⇒ (a). (a) ⇒ (b). By (2.16) of Chapter 2 applied to the positive part of the martingale of Section 2.1 (d) we have

$$P\left\{\sup_n \frac{s_n^+}{n} \geq \varepsilon\right\} \leq \varepsilon^{-1} \int_{\{\sup_n s_n/n \geq \varepsilon\}} y_1^+\, dP. \tag{4.39}$$

Hence, letting $z = \sup_n (s_n^+/n)$, we obtain

$$Ez = \int_0^\infty P\{z \geq \varepsilon\}\, d\varepsilon \leq 1 + \int_1^\infty \varepsilon^{-1} \int_{\{z \geq \varepsilon\}} y_1^+\, dP\, d\varepsilon$$

$$= 1 + \int_{\{z \geq 1\}} y_1^+ \left(\int_1^z \varepsilon^{-1}\, d\varepsilon\right) dP \leq 1 + E(y_1^+ \log^+ z).$$

It is easily verified that for all $a, b \geq 0$

$$a \log^+ b \leq a \log^+ a + e^{-1} b$$

and hence

$$Ez \leq \frac{e}{e-1}\, [1 + E(y_1^+ \log^+ y_1^+)] < \infty.$$

(a) $\Rightarrow$ (d). The proof that (a) $\Rightarrow$ (b) applied to the sequence $y_1^+$, $y_2^+, \ldots$ shows that

$$E\left[\sup_n \frac{y_n}{n}\right] \le E\left[\sup_n \frac{y_n^+}{n}\right] \le E\left[\sup_n n^{-1}\left(\sum_1^n y_k^+\right)\right] < \infty.$$

(e) $\Rightarrow$ (a). Suppose that (a) does not hold. By Theorem 4.7 it suffices to exhibit an extended stopping variable $t$ such that

$$E(y_t/t) = \infty.$$

For $c > 0$ let $t = \inf\{n: y_n \ge cn\}$. Since

$$\sum_1^\infty P\{y_n \ge cn\} = \sum_1^\infty P\{y_1 \ge cn\} < \infty,$$

it follows that for all $c$ sufficiently large

$$P\{t = \infty\} = \prod_1^\infty P\{y_n < cn\} > 0. \tag{4.40}$$

Fix $c > 1$ so large that (4.40) holds. Then

$$\begin{aligned}
E\left(\frac{y_t}{t}\right) &\ge \sum_{n=1}^\infty n^{-1}\int_{\{t=n\}} y_n = \sum_{n=1}^\infty n^{-1}P\{t \ge n\}\int_{\{y_n \ge cn\}} y_n \\
&\ge P\{t = \infty\}\sum_{n=1}^\infty n^{-1}\sum_{k=n}^\infty \int_{\{ck \le y_1 < c(k+1)\}} y_1 \\
&\ge P\{t = \infty\}\text{ const}\sum_{k=1}^\infty \log(c(k+1))\int_{\{ck \le y_1 < c(k+1)\}} y_1 \\
&\ge P\{t = \infty\}\text{ const } E(y_1^+ \log^+ y_1^+) = \infty.
\end{aligned}$$

(c) $\Rightarrow$ (a). Again suppose that (a) does not hold. Without loss of generality we may assume that $\mu = 0$, so that by the strong law of large numbers $\dfrac{s_t}{t} = 0$ on $\{t = \infty\}$ for any extended s.v. $t$. Moreover, $\dfrac{s_t}{t} \ge -\dfrac{s_{t-1}^-}{t} + \dfrac{y_t}{t}$ on $\{t < \infty\}$, and hence by the preceding part of the proof it suffices to show that

$$\int_{\{t<\infty\}}\left(\frac{s_{t-1}^-}{t}\right) < \infty$$

for the extended s.v. $t$ defined there. But

$$\int_{\{t<\infty\}} \frac{s_{t-1}^-}{t} \leq \int_{\{t<\infty\}} \left[\frac{1}{t}\sum_{k=1}^{t-1} y_k^-\right] = \sum_{n=1}^{\infty} n^{-1} \int_{\{t=n\}} \sum_{k=1}^{n-1} y_k^-$$

$$= \sum_{n=1}^{\infty} n^{-1} P\{y_n \geq cn\} \sum_{k=1}^{n-1} \int_{\{t \geq n\}} y_k^-$$

$$\leq Ey_1^- \sum_{n=1}^{\infty} P\{y_1 \geq cn\} < \infty,$$

since $E|y_1| < \infty$.

***Remark*** The inequality (4.39) shows that (4.31) does not hold for the reward sequence $\dfrac{s_n}{n}$ even when $V = \infty$.

## 9. An Application to Martingale Theory

The following application of Theorem 4.7 to martingale theory is of interest. Let $\{x_n, \mathscr{F}_n, 1 \leq n < \infty\}$ be a supermartingale with $\sup_n E|x_n| < \infty$. Theorem 2.1 says that $x_\infty = \lim_n x_n$ exists a.s. and $E|x_\infty| < \infty$.

In the terminology of the theory of optimal stopping, Lemma 3.3 (Theorem 2.2) says that under some conditions

$$v_n = Ex_n \qquad (n = 1, 2, \ldots).$$

Similarly one suspects that $\inf_t Ex_t$ ought to be $Ex_\infty$. (It follows from Theorem 2.2 that for every s.v. $t \leq n$

$$Ex_t^- \leq Ex_n^-.$$

Hence by Fatou's lemma $Ex_t^- \leq \liminf_{n\to\infty} Ex_{\min(t,n)}^- \leq \sup E|x_n| < \infty$ for every s.v. $t$.) A precise statement is

**Theorem 4.15** If $\{x_n, \mathscr{F}_n, 1 \leq n < \infty\}$ is a supermartingale such that $\sup E|x_n| < \infty$, the following are equivalent:

(a) $(x_n^-)$ is uniformly integrable,

(b) $$v_n = Ex_n \qquad (n = 1, 2. \ldots),$$

(c) $$Ex_\infty = \inf_t Ex_t.$$

***Proof*** (a) $\Rightarrow$ (b): This follows from Theorem 2.2. (b) $\Rightarrow$ (c): For each $n$, $v_n = Ex_n$ implies that $x_n \geq E(x_t \mid \mathscr{F}_n)$ for each $t \in C_n$, and hence by Theorem 4.7 that $x_n \geq E(x_\infty \mid \mathscr{F}_n)$. Thus for every s.v. $t$

$$\int_{(t=n)} x_n \geq \int_{(t=n)} x_\infty,$$

and hence (c) holds.

(c) $\Rightarrow$ (a): From Theorem 4.7, putting $\gamma_n^* = \operatorname*{ess\,inf}_{t \geq n} E(x_t \mid \mathscr{F}_n)$, we have $\gamma_n^* \leq E(x_\infty \mid \mathscr{F}_n)$ $(n = 1, 2, \ldots)$. From (c) it follows that $\gamma_n^* = E(x_\infty \mid \mathscr{F}_n)$ and hence that $x_n \geq E(x_\infty \mid \mathscr{F}_n)$ $(n = 1, 2, \ldots)$. The uniform integrability of $(x_n^-)$ follows from Lemma 2.2.

## PROBLEMS

**1.** Show that if $E(\sup x_n^+) < \infty$ and $t \in C$, there exists a $\tau \in C$ such that $\tau \leq t$ and $Ex_\tau = \sup_{\tau' \leq t} Ex_{\tau'}$. (*Hint:* Consider

$$\{x_{\min(t,n)}, \mathscr{F}_n\}_1^\infty.)$$

**2.** Use the framework of Section 6(a) to show that $\sigma$ may be *pointwise* worse than any other s.v., i.e., that for all s.v.'s $t$, $P\{x_\sigma \leq x_t\} = 1$. Where in Chapter 3 did we encounter such an example?

**3.** With $s$ defined by

$$s = \text{first } n \quad \text{such that } x_n = \gamma_n',$$

show that $s = \lim_{N \to \infty} s^N$. In the monotone case this s.v. is the "$s$" of (3.19).

**4.** Let $\{x_n, \mathscr{F}_n\}_1^\infty$ be any stochastic sequence with $E(\sup x_n^+) < \infty$ (but do not assume $Ex_n^- < \infty$). Prove Theorem 4.1 from first principles. Now assume that $E(x_n^-) < \infty$ $(n = 1, 2, \ldots)$ and use Lemma 4.12 to remove the condition $E(\sup x_n^+) < \infty$.

**5.** *Continuation.* What other results of Chapter 4 do not require the hypothesis $E(x_n^-) < \infty$?

**6.** We say that a supermartingale $\{y_n, \mathscr{F}_n, 1 \leq n < \infty\}$ is semiregular if

$$E(y_t \mid \mathscr{F}_n) \leq y_n \quad \text{on} \quad \{t \geq n\}$$

for every s.v. $t$ for which $Ey_t$ exists. Show that if $Ex_t$ exists for every s.v. $t$, then $\{\gamma_n, \mathscr{F}_n, 1 \le n < \infty\}$ is semi-regular. (*Hint:* Observe that it suffices to show

$$E(\gamma_t \mid \mathscr{F}_n) \le \gamma_n \quad \text{on} \quad \{t \ge n, \gamma_n \le b\}$$

and let $b \to \infty$. Now modify appropriately the proof of Theorem 4.6.)

**7.** Let $\{y_n, \mathscr{F}_n, 1 \le n < \infty\}$ be a supermartingale and $y_\infty$ *any r.v.* For any extended s.v. $t$ we agree that $y_t = y_\infty$ on $\{t = \infty\}$. (Note that this differs from our usual convention.) Show that if there exists an *extended* s.v. $t$ such that $Ey_t^- < \infty$ and

$$(^*) \qquad E(y_t \mid \mathscr{F}_n) \le y_n \quad \text{on} \quad \{t \ge n\} \qquad \text{for all } n = 1, 2, \ldots,$$

then (*) continues to hold with $t$ replaced by any *s.v.* $\tau \le t$. (*Application:* If $V < \infty$ and $\sigma$ is optimal in $\bar{C}$, then for each $\varepsilon > 0$

$$\sigma(\varepsilon) = \text{first } n \quad \text{such that} \quad x_n \ge \gamma_n - \varepsilon$$

is "$\varepsilon$-optimal" in $C$ (i.e., $\sigma(\varepsilon) \in C$ and $Ex_{\sigma(\varepsilon)} \ge V - \varepsilon$).)

**8.** Let $y_1, y_2, \ldots$ be i.i.d. with $Ey_1 < 0$, $E[(y_1^+)^2] = \infty$; let

$$\begin{aligned} t &= \text{first } n \ge 1 \quad \text{such that} \quad \sum_1^n y_k > 0 \\ &= \infty \quad \text{if no such } n \text{ exists.} \end{aligned}$$

Show that $E\left(\sum_1^t y_k\right)^+ = \infty$. (*Hint:* Apply Theorem 4.13.)

**9.** Let $\mathscr{F}_1 \subset \mathscr{F}_2 \subset \cdots$ and let $-x_n = y_n + c_n$, where $y_n$ and $c_n$ are $\mathscr{F}_n$-measurable, $y_n \ge 0$, and $0 \le c_n \uparrow \infty$. Define

$$\begin{aligned} \tilde{\gamma}_N^N &= -c_N \\ \tilde{\gamma}_N^N &= \max(x_n, E(\tilde{\gamma}_{n+1}^N \mid \mathscr{F}_n)) \\ & \qquad\qquad (n = N-1, \ldots, 1). \end{aligned}$$

Show that $\tilde{\gamma}_n^N \downarrow \gamma_n$ as $N \uparrow \infty$. This provides an alternative method of computing the sequence $(\gamma_n)$ in many problems of interest in statistics.

**10.** In the notation of problem 2.1 with $p = \frac{1}{2}$, let $\mathscr{F}_n = \mathscr{B}(y_1, \ldots, y_n)$ and $x_n = f(s_{\min(t,n)})$, where $f$ is any real-valued function defined on $\{-a, -a+1, \ldots, b-1, b\}$. Let $g$ denote the smallest concave function majorizing $f$. Show that $\gamma_n = g(s_{\min(t,n)})$. (See Dynkin and Yushkevich [1].)

# Chapter 5

# The Markov and Independent Cases

In examples we frequently start with a sequence $(y_n)$ of *independent* random variables, $\mathscr{F}_n = \mathscr{B}(y_1, \ldots, y_n)$, and take $x_n$ to be some function of $y_n$, e.g., the secretary problem and the parking problem. In such cases it is easy to see by backward induction (putting $\mathscr{F}_0 = \{\phi, \Omega\}$) that

$$E(\gamma_n^N \mid \mathscr{F}_{n-1}) = E\gamma_n^N = v_n^N \qquad (n = N, \ldots, 1). \tag{5.1}$$

Defining $v_{N+1}^N = -\infty$, we have

$$\gamma_n^N = \max(x_n, v_{n+1}^N) \qquad (n = N, \ldots, 1) \tag{5.2}$$

so that $\gamma_1^N, \ldots, \gamma_N^N$ are independent and

$$s^N = \text{first } n \geq 1 \quad \text{such that} \quad x_n \geq v_{n+1}^N.$$

Now letting $N \to \infty$, we find

$$\gamma_n' = \max(x_n, v_{n+1}') \qquad (n = 1, 2, \ldots), \tag{5.3}$$

where $\gamma_n'$ and $v_n'$ are defined by (4.8), so in those cases in which $(\gamma_n') = (\gamma_n)$ we have

$$\gamma_n = \max(x_n, v_{n+1}) \qquad (n = 1, 2, \ldots), \tag{5.4}$$

$$\gamma_1, \gamma_2, \ldots \text{ are independent,}$$

and

$$\sigma = \text{first } n \geq 1 \quad \text{such that} \quad x_n \geq v_{n+1}.$$

Thus from the assumption that $(\gamma_n) = (\gamma'_n)$ and the fact that $(\gamma'_n)$ is *in principle* computable by backward induction and passing to the limit we have been able to draw some important conclusions. It is this observation which suggests possible applications of Theorem 4.8, and in this chapter we define the Markov case and the independent case and show how Theorem 4.8 allows us to derive useful qualitative properties of the sequence $(\gamma_n)$ and the rule $\sigma$.

## 1. The Markov Case–Definition and Basic Theorems

**Definition** Let $\{x_n, \mathscr{F}_n\}_1^\infty$ be a stochastic sequence. Suppose that for each $n = 1, 2, \ldots$ there exists a measurable space $(Z_n, \mathscr{Z}_n)$ and an $\mathscr{F}_n$-measurable random variable $z_n$ taking values in $Z_n$ such that $x_n$ can be expressed as some $\mathscr{Z}_n$-measurable function, say $\varphi_n$, of $z_n$. We say that this provides a *Markov representation* of the sequence $\{x_n, \mathscr{F}_n\}_1^\infty$ if

(5.5) $$P\{z_{n+1} \in B \mid \mathscr{F}_n\} = P\{z_{n+1} \in B \mid z_n\} \qquad (B \in \mathscr{Z}_{n+1}, n = 1, 2, \ldots).$$

If in addition $Z_1 = Z_2 = \cdots = Z$, $\mathscr{Z}_1 = \mathscr{Z}_2 = \cdots = \mathscr{Z}$, and $P\{z_{n+1} \in B \mid z_n = z\}$ is for each $n$ and $B \in \mathscr{Z}$ a function on $Z$ which does not depend on $n$, then we have a *stationary* Markov representation of $\{x_n, \mathscr{F}_n\}_1^\infty$.

### Examples

(a) Consider the problem of $x_n = \max(y_1, \ldots, y_n) - c_n$ (Section 3.6(a)). Putting $z_n = x_n$ we obtain a Markov representation; putting $z_n = \max(y_1, \ldots, y_n)$ we obtain a stationary Markov representation. A similar result holds for the reward sequence $\frac{s_n}{n}$ (Section 4.7).

(b) For the problem of Section 3.1(h) we obtain a stationary Markov representation by putting $z_n = \pi_n$. Note that for this problem we cannot put $z_n = h(\pi_n)$ or $z_n = x_n$.

Our first result states that in the Markov case we may with no loss of generality assume that

(i) the underlying probability space is $(\Omega', \mathscr{F}', P')$, where $\Omega' = Z_1 \times Z_2 \times \cdots$, $\mathscr{F}' = \mathscr{Z}_1 \times \mathscr{Z}_2 \times \cdots$, and $P'$ is the image of $P$ under $(z_1, z_2, \ldots)$;

(ii) for each $n$, $\mathscr{F}'_n = \mathscr{B}(z_1, \ldots, z_n)$, where $z_n$ is now regarded as the $n$th coordinate variable on $\Omega'$;

(iii) a stopping rule $t$ is defined by a sequence of absorbing sets $(B_n)$ in the sense that for each $n = 1, 2, \ldots, B_n \in \mathscr{Z}_n$ and $t =$ first $n \geq 1$ such that $z_n \in B_n$.

We shall state these results more precisely in Theorem 5.1 below, the proof of which depends on two lemmas.

**Lemma 5.1** In the Markov case $\gamma_n$ is measurable with respect to $\mathscr{B}(z_n)$ $(n = 1, 2, \ldots)$.

***Proof*** The proof is an immediate application of Theorem 4.8 and the definition of the Markov case.

**Lemma 5.2** For each $b > 0$ let $(x_n(b))$ and $(\gamma_n(b))$ be defined as in Section 4.5 and for any $n = 1, 2, \ldots$ put $t(b) =$ first $k \geq n$ such that $x_k \geq \gamma_k(b) - b^{-1}$. Then $t(b) \in C_n$ for all $b$ and

$$\gamma_n = \sup_b E(x_{t(b)} \mid \mathscr{F}_n), \; v_n = \sup_b Ex_{t(b)}.$$

***Proof*** The lemma follows at once from Lemmas 4.8, 4.12, and 4.13. Alternatively Lemmas 4.10, 4.12 and the proof of Lemma 4.9 supply the basic ingredients. (See also problem 7 of Chapter 4.)

**Definition** Suppose that we are in the Markov case. Denote by $D_n(\overline{D}_n)$ the subclass of $C_n(\overline{C}_n)$ with the property that for each $t \in D_n(\overline{D}_n)$ there exist $B_k \in \mathscr{L}_k$, $k = n, n+1, \ldots$ such that

$$\begin{aligned} t &= \text{first } k \geq n \quad \text{such that} \quad z_k \in B_k \\ &= \infty \quad \text{if no such } k \text{ exists.} \end{aligned}$$

The reader may regard Theorem 5.1 as being intuitively obvious, since in the Markov case it seems clear that the decision to stop should be based on the present state of the system without reference to the states previously occupied. This principle is often taken for granted in treating optimal stopping problems. Its rigorous justification is closely related to the discussion of randomized stopping rules in Section 5.3 below.

**Theorem 5.1** In the Markov case for each $n = 1, 2, \ldots$

$$\sigma_n \in \overline{D}_n,$$

and

$$\gamma_n = \operatorname*{ess\,sup}_{t \in D_n} E(x_t \mid \mathscr{F}_n) = \operatorname*{ess\,sup}_{t \in D_n} E(x_t \mid z_n),$$

$$v_n = \sup_{D_n} Ex_t.$$

***Proof*** Theorem 5.1 follows at once from Lemmas 5.1 and 5.2.

In the following we are interested in the stationary Markov case. We may and do assume that the underlying probability space has the structure discussed above. We also assume that there exists a transition probability governing the evolution of the sequence $(z_n)$, i.e., that there exists a real-valued function $q(\cdot, \cdot)$ on $Z \times \mathscr{Z}$ such that $q(\cdot, B)$ is $\mathscr{Z}$-measurable for each $B \in \mathscr{Z}$, $q(z, \cdot)$ is a probability on $\mathscr{Z}$ for each $z \in Z$, and $q(z_n, B)$ is a version of $P\{z_{n+1} \in B \mid z_n\}$ $(n = 1, 2, \ldots)$. Since by Lemma 5.1 $\gamma_n$ ($\gamma_n^N$, etc.) is $\mathscr{B}(z_n)$-measurable we shall write $\gamma_n(z)$ ($\gamma_n^N(z)$, etc.) to denote the value at $z$ of that (measurable) function on $Z$ which equals $\gamma_n(\omega)$ ($\gamma_n^N(\omega)$, etc.) when $z_n(\omega) = z$. Then (cf. Theorem 3.2)

$$\gamma_N^N(z) = \varphi_N(z)$$

(5.6)

$$\gamma_n^N(z) = \max\left(\varphi_n(z), \int_Z q(z, dz')\gamma_{n+1}^N(z')\right) \qquad (n = N-1, \ldots, 1).$$

We denote by $P^z$ the probability on $\mathscr{F}$ governing the behavior of $(z_n)$ starting from the initial point $z$, and for each $n = 0, 1, 2, \ldots$ we let $V_n(z)$ denote the value (in the sense of Section 3.1) of $\{\varphi_{n+k}(z_k), \mathscr{F}_k\}_1^\infty$ on $(\Omega, \mathscr{F}, P^z)$ and set $\Gamma_n(z) = \max(\varphi_n(z), V_n(z))$ $(n = 1, 2, \ldots)$. (We have tacitly assumed that for each $n = 0, 1, \ldots$ and $z \in Z$ $\{\varphi_{n+k}(z_k), \mathscr{F}_k\}_1^\infty$ is an integrable stochastic sequence on $(\Omega, \mathscr{F}, P^z)$. Since

$$E^z|\varphi_{n+k}(z_k)| = E(|\varphi_{n+k}(z_{n+k})| \| z_n = z),$$

and $\{\varphi_n(z_n), \mathscr{F}_n\}_1^\infty$ is an integrable stochastic sequence on $(\Omega, \mathscr{F}, P)$ it follows that for each $n = 1, 2, \ldots$ and a.e. (with respect to $P\{z_n \in (\cdot)\}$) $z$, $E^z|\varphi_{n+k}(z_k)| < \infty$. Our additional assumption should cause no trouble in practice.)

**Theorem 5.2** In the stationary Markov case there is a version of $(\gamma_n)$ such that for each $n = 1, 2, \ldots$

$$V_n(\cdot) = E(\gamma_{n+1}(z_{n+1}) \mid z_n = \cdot)$$

$$\Gamma_n(\cdot) = \gamma_n(\cdot).$$

***Proof*** The version of $(\gamma_n)$ in which we are interested is the triple limit of Theorem 4.8. It suffices to prove the statement regarding $V_n$, and by Theorem 4.8 we may assume that $|\varphi_n| \le B < \infty$, for all $n$. If $V_n^N(\cdot)$ is defined analogously to $V_n(\cdot)$ relative to the class of rules which take at most $N - n$ observations it suffices to show for $n = 1, 2, \ldots$

$$V_n^N(z) = E(\gamma_{n+1}^N(z_{n+1}) \mid z_n = z) \qquad (z \in Z, N = n+1, \ldots). \tag{5.7}$$

It is easy to see using (5.6) that for $N = 2, 3, \ldots$ (5.7) holds for $n = N - 1, \ldots, 1$.

**Corollary** Under the assumptions of Theorem 5.2 suppose that $\varphi_n = \varphi - n$ $(n = 1, 2, \ldots)$ and let $B = \{z\colon \varphi(z) \geq V_0(z)\}$. Then for each $n = 1, 2, \ldots$

$$\{\sigma = n\} = \{z_1 \notin B, \ldots, z_{n-1} \notin B, z_n \in B\}.$$

***Proof*** Obviously $V_n = V_0 - n$ $(n = 0, 1, \ldots)$. The corollary follows at once from the theorem and the definition of $\sigma$.

***Remark*** By straightforward but tedious arguments we can prove a similar result if $x_n = \sum_1^{n-1} w_k(z_k) + \varphi_n(z_n)$, where $z_1, z_2, \ldots$ have the (stationary) Markov property (5.5). Putting

$$V_n(z) = \sup_t E^z\left(\sum_1^{t-1} w_{n+k}(z_k) + \varphi_{n+t}(z_t)\right)$$

we now have

$$E(\gamma_{n+1} \mid z_1, \ldots, z_n) = \sum_1^{n-1} w_k(z_k) + V_n(z_n) \qquad (n = 1, 2, \ldots)$$

and hence

$$\begin{aligned} \sigma &= \text{first } n \geq 1 \quad \text{such that} \quad \varphi_n(z_n) \geq V_n(z_n) \\ &= \infty \quad \text{if no such } n \text{ exists.} \end{aligned}$$

This remark will be used in Section 5.2(b).

## 2. The Markov Case–Applications

(a) A Bayes solution to the problem of testing a simple hypothesis against a simple alternative with constant cost and independent, identically distributed observations is a Wald sequential probability ratio test (see Example 3.1(h)). By Theorem 4.5 $\sigma$ is optimal, and the corollary to Theorem 5.2 provides a rigorous foundation for the heuristic arguments presented in Chapter 3 where the problem was introduced.

***Remark*** Although Theorem 5.1 implies that we may in this problem disregard the random variables $y_1, y_2, \ldots$ and pretend that we "observe" $\pi_0, \pi_1, \pi_2, \ldots$, nevertheless, the $y$'s play a part in our analysis. Implicit in our use of the representation (see (3.3))

$$V_0(\pi) = \sup_t \{-[\pi(a\alpha_0 + E_0 t) + (1 - \pi)(b\alpha_1 + E_1 t)]\}$$

is the assumption that we are originally dealing with s.v.'s $t$ defined in terms of $y_1, y_2, \ldots$. It is this assumption that allows us to reason that $\alpha_1$ and $E_i(t)$ $(i = 0, 1)$ do not depend on $\pi_0 = \pi$ and hence that $V_0(\cdot)$ is convex.

(b) In addition to the assumptions of Section 4.3(c) suppose that $r_k = p^{k-1}q$ $(k = 1, 2, \ldots)$, where $0 < p, q < 1, p + q = 1$. Then there exists a number $\pi^*$ $(0 < \pi^* < 1)$ such that

$$\sigma = \text{first } n \geq 0 \quad \text{such that} \quad \pi_n \geq \pi^*.$$

***Proof*** Writing

$$\begin{aligned} x_n &= -\left[c(1 - \pi_n) + \sum_{k=0}^{n-1} (n - k)p_k^n\right] \\ &= -\left[c(1 - \pi_n) + \sum_0^{n-1} \pi_k^n\right] \\ &= -\left[c(1 - \pi_n) + \sum_0^{n-1} \pi_k + \sum_0^{n-1} (\pi_k^n - \pi_k)\right], \end{aligned}$$

where we have put $\pi_k^n = \sum_{i=0}^{k} p_i^n$ $(n = 0, 1, \ldots, k = 0, 1, \ldots)$, we observe that

$$\left\{f_n \equiv \sum_0^{n-1} (\pi_k^n - \pi_k), \mathscr{F}_n, 0 \leq n < \infty\right\}$$

is a $C$-regular martingale. Straightforward calculations show it is a martingale with $Ef_0 = 0$. Moreover, if $t \in C$, then by Lemma 3.3 (Theorem 2.2) applied to the bounded martingale

$$\{\pi_k^n, \mathscr{F}_n, k \leq n < \infty\}$$

we have

$$\begin{aligned} \infty > E\left(\sum_{k=0}^{t-1} \pi_k^t\right) &= \sum_{n=0}^{\infty} \int_{(t=n)} \sum_{k=0}^{n-1} \pi_k^n = \sum_{k=0}^{\infty} \sum_{n=k+1}^{\infty} \int_{(t=n)} \pi_k^n \\ &= \sum_{k=0}^{\infty} \int_{(t>k)} \pi_k^t = \sum \int_{(t>k)} \pi_k = E\left(\sum_0^{t-1} \pi_k\right), \end{aligned}$$

where the last equality follows by reversing the steps which led to the first three. Hence

$$\int_{(t>n)} |f_n| \le \int_{(t>n)} \sum_0^{n-1} (\pi_k^n + \pi_k) = \int_{(t>n)} \sum_0^{n-1} (\pi_k^t + \pi_k)$$
$$\le \int_{(t>n)} \sum_0^{t-1} (\pi_k^t + \pi_k) \to 0 \quad \text{as} \quad n \to \infty,$$

and it follows from Lemma 3.3 (Theorem 2.2) that

$$\{f_n, \mathscr{F}_n, 0 \le n < \infty\}$$

is $C$-regular. Since $Ef_t = Ef_0 = 0$ $(t \in C)$ we may and do assume that

$$x_n = -\left[c(1 - \pi_n) + \sum_0^{n-1} \pi_k\right].$$

Writing $V_0(\pi) = \sup_{t \ge 1} E^\pi\left[-c(1 - \pi_t) - \sum_{k=0}^{t-1} \pi_k\right]$, where $E^\pi$ denotes expectation under the assumption that $\pi_0 = \pi$, it is easily seen from the conditions imposed on $(r_k)$ and the remark at the end of Section 5.1 that

$$\begin{aligned} \sigma &= \text{first } n \ge 0 \quad \text{such that} \quad -c(1 - \pi_n) \ge V_0(\pi_n) \\ &= \text{first } n \ge 0 \quad \text{such that} \quad \pi_n - 1 \ge c^{-1}V_0(\pi_n). \end{aligned}$$

From the representation

$$-V_0(\pi) = \inf_{t \ge 1} \left\{\pi E_1 t + (1 - \pi)q \sum_{n=1}^{\infty} p^{n-1}(cP_0\{t < n\} + \hat{E}((t - n)^+ \mid \theta = n))\right\}$$

(see the remark following (a) above), where $\hat{E}((t - n)^+ \mid \theta = n) = \sum_{n+1}^{\infty} (k - n) \int_{(t \ge n)} P_1(t = k \mid \mathscr{F}_{n-1})\, dP_0$, we see that $V_0(\cdot)$ is convex. Since $V_0(1) = -1$, the result follows.

(c) Let $y_1, y_2, \ldots$ be a sequence of i.i.d. random variables. The $y$'s have probability density function $f$ with respect to some $\sigma$-finite measure $\mu$, and it is known that $f$ is one of two specified density functions, $f_0$ or $f_1$. If $f = f_1$, the observations each cost a unit amount. If $f = f_0$ there is no sampling cost, but we incur a unit cost if we ever stop sampling. Hence if it appears that we are sampling from $f_1$, we would like to stop as soon as possible, whereas we would like to continue sampling indefinitely as long as it appears that $f = f_0$. If there is an a priori probability $\pi$ $(0 < \pi < 1)$ that $f = f_0$, we would like to find an extended s.v. minimizing

$$\pi P_0\{t < \infty\} + \bar{\pi} E_1 t, \tag{5.8}$$

where $\bar{\pi} = 1 - \pi$ and we write $P_i(E_i)$ to denote the probability (expectation) on the space of infinite sequences $(y_1, y_2, \ldots)$ determined by $f_i$ $(i = 0, 1)$.

Let $f_{in} = f_i(y_1) \cdots f_i(y_n)$ $(i = 0, 1; n = 1, 2, \ldots)$. For any stopping rule $t$ for which $P_0(t < \infty, f_{1t} = 0) = 0$ and $E_1 t < \infty$ (in trying to minimize (5.8) it suffices to consider such rules), omitting the terms $d\mu(y_1) \cdots d\mu(y_n)$, we have

$$P_0(t < \infty) = \sum_1^\infty \int_{(t=n)} f_{0n} = \sum_1^\infty \int_{(t=n)} \frac{f_{0n}}{f_{1n}} \cdot f_{1n} = E_1\left(\frac{f_{0t}}{f_{1t}}\right).$$

Putting $-x_n = \pi \dfrac{f_{0n}}{f_{1n}} + \bar{\pi} n$, $\mathscr{F}_n = \mathscr{B}(y_1, \ldots, y_n)$ $(n \geq 1)$, we see that to minimize (5.8) it suffices to maximize $E_1 x_t$, i.e., to solve the optimal stopping problem for $\{x_n, \mathscr{F}_n\}_1^\infty$ under the probability $P_1$.

Let $z_n = \dfrac{f_{0n}}{f_{1n}}$ $(n \geq 1)$. We are in the stationary Markov case. By Theorem 4.5 $\sigma$ is optimal in $C$, and by Theorem 5.2

$$\sigma = \text{first } n \geq 1 \quad \text{such that} \quad -\pi z_n \geq V_0(z_n).$$

It is easy to see that $V_0(0) = -\bar{\pi}$ and that $V_0(\cdot)$ is convex. It follows that for some constant $c$

$$\sigma = \text{first } n \geq 1 \quad \text{such that} \quad \frac{f_{0n}}{f_{1n}} \leq c. \tag{5.9}$$

However, since $\{z_n, \mathscr{F}_n\}_1^\infty$ is a martingale under $P_1$, we have $s^N \equiv 1$ for all $N$ and hence $V^N \nrightarrow V$. Putting $s_n = \sum_1^n \log \dfrac{f_0(y_k)}{f_1(y_k)}$, we note that (5.9) becomes

$$\sigma = \text{first } n \geq 1 \quad \text{such that} \quad s_n \leq \log c,$$

and (provided $P_0(\sigma = 1) < 1$)

$$1 > P_0(\sigma < \infty) = E_1\left(\frac{f_{0\sigma}}{f_{1\sigma}}\right) = E_1(e^{s_\sigma}),$$

in agreement with the results of Section 2.5. To compute $(\gamma_n)$ we set

$$\hat{\gamma}_N^N = -\bar{\pi} N$$

$$\hat{\gamma}_n^N = \max(x_n, E(\hat{\gamma}_{n+1}^N \mid \mathscr{F}_n)) \qquad (n = N-1, \ldots, 1).$$

Then by Lemma 4.9 $\lim_{N\to\infty} \hat{\gamma}_n^N = \gamma_n \ (n \geq 1)$.

(d) The following alternative formulation of (c) provides an example of a problem in which extended stopping rules are of interest but the reward $x_\infty$ at time $+\infty$ appropriate to the problem is *not* $\lim\sup_{n\to\infty} x_n$. Hence the results of Chapter 4 do not apply. For example, Theorem 4.7, which implies that $V = \sup_t Ex_t$ is the same regardless of whether the sup is taken over the class of stopping rules or the class of extended stopping rules, is not true. Nevertheless, an appropriate extension of our general theory can be carried out in a straightforward manner which we indicate below.

Consider the same problem as in (c) above, but let $P = \pi P_0 + \bar{\pi} P_1$. Under $P$ the joint density of $y_1, \ldots, y_n$ is for each $n$ equal to $\pi f_{0n} + \bar{\pi} f_{1n}$. Set $\pi_n = \dfrac{\pi f_{0n}}{\pi f_{0n} + \bar{\pi} f_{1n}}$, $\bar{\pi}_n = 1 - \pi_n$. Hence

$$\begin{aligned}(5.10) \qquad \pi P_0(t < \infty) + \bar{\pi} E_1 t &= \sum_1^\infty \int_{(t=n)} (\pi f_{0n} + \bar{\pi} n f_{1n}) \\ &= \sum_1^\infty \int_{(t=n)} (\pi_n + n\bar{\pi}_n)(\pi f_{0n} + \bar{\pi} f_{1n}) \\ &= \int_{(t<\infty)} (\pi_t + t\bar{\pi}_t)\, dP.\end{aligned}$$

Since by Section 2.2(d)

$$P_1\left\{\frac{f_{0n}}{f_{1n}} \to 0\right\} = 1 \quad \text{and} \quad P_0\left\{\frac{f_{1n}}{f_{0n}} \to 0\right\} = 1,$$

it follows that $\pi_\infty = \lim_{n\to\infty} \pi_n$ exists a.s. $P$ and $P_0(\pi_\infty = 1) = 1$, $P_1(\pi_\infty = 1) = 0$. Putting

$$-x_n = \pi_n + n\bar{\pi}_n, \quad \mathscr{F}_n = \mathscr{B}(y_1, \ldots, y_n) \qquad (n \geq 1)$$

$$-x_\infty = \infty I_{\{\bar{\pi}_\infty = 1\}},$$

and agreeing that for any extended s.v. $t$

$$x_t = \begin{cases} x_n & \text{if } t = n \\ x_\infty & \text{if } t = \infty, \end{cases} \qquad (n = 1, 2, \ldots)$$

we have from (5.10)

$$\pi P_0(t < \infty) + \bar{\pi} E_1 t = -Ex_t.$$

Thus minimizing (5.8) is equivalent to maximizing $Ex_t$. This latter problem, however, does not fall within the scope of the theory of optimal stopping as we have developed it, for $P\{x_\infty \neq \limsup x_n\} > 0$. Furthermore, it is easy to see that $\{\pi_n, \mathscr{F}_n\}_1^\infty$ is a bounded martingale; and hence by Lemma 3.3 (Theorem 2.2)

$$Ex_t = -\pi - E_1 t$$

for every s.v. $t$ with $P\{t < \infty\} = 1$.

More generally, suppose then that $\{x_n, \mathscr{F}_n\}_1^\infty$ is *any* integrable stochastic sequence and $x_\infty$ an *arbitrary* $\mathscr{B}\left(\bigcup_1^\infty \mathscr{F}_n\right)$-measurable r.v. For any extended s.v. $t$ let $x_t$ be $x_n$ on $\{t = n\}$ for $n = 1, 2, \ldots$ and $x_\infty$ on $\{t = \infty\}$. Let $\bar{C}$ denote the class of all those $t$ for which $Ex_t^- < \infty$, $\bar{V} = \sup_{t \in \bar{C}} Ex_t$, etc. An appropriate extension of our general theory may be carried out in a straightforward fashion. For example, the fundamental Lemma 4.11 (a) becomes:

**Lemma** Define

$$\tilde{\gamma}_N^N = E(\max(x_\infty, \sup_{k \geq N} x_k) \mid \mathscr{F}_N)$$

$$\tilde{\gamma}_n^N = \max(x_n, E(\tilde{\gamma}_{n+1}^N \mid \mathscr{F}_n)) \qquad (n = N-1, \ldots, 1),$$

$$\tilde{\gamma}_n = \lim_{N \to \infty} \tilde{\gamma}_n^N.$$

If $E(\max[x_\infty, \sup_n x_n])^+ < \infty$, then $(\tilde{\gamma}_n) = (\bar{\gamma}_n)$. If in addition $x_\infty \leq \limsup_{n \to \infty} x_n$, then $(\bar{\gamma}_n) = (\gamma_n)$.

We omit the details. (See problems 8–10.)

## 3. Randomized Stopping Rules

From one point of view Theorem 5.1 is a result on the elimination of "randomized" stopping rules. In the Markov case at any time $n$ the conditional joint distribution of $x_n, x_{n+1}, \ldots$ depends on the past only through the value of $z_n$, and thus it seems reasonable that our decision to stop at stage $n$ should depend only on $z_n$. Allowing our decision to depend on previous observations is tantamount to basing our decision on an auxiliary "irrelevant" randomization.

Now suppose that there is given an integrable stochastic sequence $\{x_n, \mathscr{F}_n\}_1^\infty$ with the property that there exists an underlying sequence $y_1, y_2, \ldots$ of (in general abstract-valued) random variables such that $\mathscr{F}_n = \mathscr{B}(y_1, \ldots, y_n)$ $(n = 1, 2, \ldots)$. Then it is easy to see that the optimal stopping problem for $\{x_n, \mathscr{F}_n\}_1^\infty$ is the same regardless of the specific underlying probability space $(\Omega, \mathscr{F}, P)$ on which $\{x_n, \mathscr{F}_n\}_1^\infty$ is defined, provided only that $(\Omega, \mathscr{F}, P)$ is consistent with the (given) joint distribution of $y_1, y_2, \ldots$.

If $\mathscr{G}_1, \mathscr{G}_2, \ldots$ is a non-decreasing sequence of sub-$\sigma$-algebras of $\mathscr{F}$ such that for each $n = 1, 2, \ldots$

(a) $$\mathscr{F}_n \subset \mathscr{G}_n$$

and

(b) $$P(A \mid \mathscr{G}_n) = P(A \mid \mathscr{F}_n) \qquad \left(A \in \mathscr{B}\left(\bigcup_1^\infty \mathscr{F}_k\right)\right),$$

then we call any s.v. $t$ relative to $(\mathscr{G}_n)$ (i.e., such that $\{t = n\} \in \mathscr{G}_n$ for all $n = 1, 2, \ldots$) a randomized s.v. for $\{x_n, \mathscr{F}_n\}_1^\infty$; and the class of randomized s.v.'s for $\{x_n, \mathscr{F}_n\}_1^\infty$ is the class of s.v.'s which can be obtained from such sequences $(\mathscr{G}_n)$ as $(\Omega, \mathscr{F}, P)$ ranges over the class of probability spaces consistent with the specified joint distribution of $y_1, y_2, \ldots$. Note that the intuitive method of randomization whereby at stage $n$ one performs an auxiliary random experiment depending (measurably) on $y_1, \ldots, y_n$ in order to decide whether to stop fits into the above scheme. In fact, in this case we would have $\mathscr{G}_n = \mathscr{B}(y_1, \ldots, y_n$, and all random experiments up to and including stage $n)$.

### Example

Let $y_k = k, x_k = y_k$ $(k = 1, 2, \ldots)$. The "natural" probability space for this situation is a space $\Omega$ consisting of a single element $\omega$. $V = \infty$ and no optimal rule exists. If we let $(\Omega, \mathscr{F}, P)$ be the unit interval with the Borel sets and Lebesgue measure, then $V = \infty$, and putting $\mathscr{G}_1 = \mathscr{G}_2 = \cdots = \mathscr{F}$, $t(\omega) = 2^k$ if $\dfrac{1}{2^k} < \omega \le \dfrac{1}{2^{k-1}}$ $(k = 1, 2, \ldots)$, we have $Ex_t = \sum_{k=1}^\infty 2^k . \dfrac{1}{2^k} = \infty$, so there is an optimal randomized s.v.

Our fundamental result is that *the introduction of randomization does not change the value, V,* of $\{x_n, \mathscr{F}_n\}_1^\infty$.

**Theorem 5.3** Let $\{x_n, \mathscr{F}_n\}_1^\infty$ be any stochastic sequence and $\mathscr{G}_1 \subset \mathscr{G}_2 \subset \cdots$ any sequence of sub-$\sigma$-algebras of $\mathscr{F}$ having properties (a) and (b). Then the $\gamma$'s for the sequence $\{x_n, \mathscr{G}_n\}_1^\infty$ are the same as the $\gamma$'s for the sequence $\{x_n, \mathscr{F}_n\}_1^\infty$.

***Proof*** The proof follows at once from Theorem 4.8 and condition (b) above.

Since the extended s.v. $\sigma$ is defined in terms of the sequence $(\gamma_n)$, Theorem 5.3 in conjunction with Theorem 4.2 implies that if $V < \infty$ we may restrict our consideration to nonrandomized rules. The first example in this section shows that if $V = \infty$ there *may* exist an optimal randomized rule but no optimal nonrandomized rule. More generally, whenever $V = \infty$, there exist s.v.'s $(t_k)$ such that $\lim_{k\to\infty} Ex_{t_k} = \infty$. By passing to a subsequence if necessary we may assume that $Ex_{t_k} \geq 2^k$; and by using the s.v. $t_k$ with probability $2^{-k}$, we might hope to obtain a randomized s.v. $t$ such that $Ex_t = \sum_{k=1}^{\infty} 2^{-k}E(x_{t_k}) \geq \sum_1^{\infty} 1 = \infty$. That this line of reasoning is *not* valid in general is the content of the first example below.

**Examples**

With the set-up of Section 4.6(a), assume that

$$a_k = k^2(k+2)$$
$$b_k = 2k^2(k-1).$$

Then $Ex_{t_n} = n + 2 \uparrow \infty$, but

$$1 \geq \frac{n+2}{n^2+n-2} = \frac{Ex_{t_n}}{Ex_{t_n}^-} \to 0.$$

Thus any mixture $(p_n)$ of the rules $(t_n)$ which formally gives $\sum_2^{\infty} p_n Ex_{t_n} = \infty$ must in fact define a rule $t$ such that $Ex_t^+ = Ex_t^- = \infty$.

The following example provides an interesting application of the idea of randomization. Let $y, y_1, y_2, \ldots$ be i.i.d., $P\{y = 1\} = p = 1 - P\{y = 0\}$, $\mathcal{F}_n = B(y_1, \ldots, y_n)$, $x_n = \dfrac{y_1 + \cdots + y_n}{n}$. We shall show that $V = V(p)$ is increasing and continuous in $p$. Let $Y_1, Y_2, \ldots$ be independent and uniform on (0, 1),

$$\mathcal{G}_n = \mathcal{B}(Y_1, \ldots, Y_n),$$

$x_n(p)$ = proportion of terms among $Y_1, \ldots, Y_n$ which are $\leq p$. By Theorem 5.3 the value of $\{x_n, \mathcal{F}_n\}_1^{\infty}$ is the same as that of $\{x_n(p), \mathcal{G}_n\}_1^{\infty}$. Since $x_n(\cdot)$ is increasing in $p$, it follows that $V(\cdot)$ is increasing in $p$. The continuity of $V(\cdot)$ follows from Theorem 4.9.

## 4. A Problem of G. Elfving

Suppose that a man owns some commodity, e.g., a house, which is for sale. From time to time he receives offers which he must accept or reject. The longer he postpones selling (waiting for a large offer) the more he loses in interest, depreciation, taxes, etc. What should he do?

We shall consider the following simple model of the above situation. Let $\{N(u), u \geq 0\}$ be a non-homogeneous Poisson process with intensity function $p(u)$. (The process has independent increments and the probability that $N(u + \delta) - N(u) = 1$ is $\delta p(u) + o(\delta)$.) Let $\tau_1, \tau_2, \ldots$ denote the time points of successive jumps in the sample paths of $\{N(u), u \geq 0\}$, i.e., with $\tau_0 = 0$

$$\tau_k = \inf \{u: u \geq \tau_{k-1}, N(u) \neq N(\tau_{k-1})\}.$$

With $\tau_1, \tau_2, \ldots$ (the times of successive offers) are associated independent non-negative random variables $y_1, y_2, \ldots$ (the amounts offered) with a common distribution $F$ and finite mean $\mu$. Furthermore, there is given a non-negative, non-increasing, right continuous discount function $r(\cdot)$ with $r(0) = 1$. For $n = 1, 2, \ldots$ let $\mathscr{H}_n = \mathscr{B}(\tau_1, \ldots, \tau_n, y_1, \ldots, y_n)$ and $x_n = y_n r(\tau_n)$. We shall assume

$$\int_0^\infty r(u) p(u)\, du < \infty,$$

so

$$E\left(\sum_1^\infty y_n r(\tau_n)\right) = \mu \int_0^\infty r(u) p(u)\, du < \infty. \tag{5.11}$$

Hence $E(\sup x_n) < \infty$ and $x_\infty = 0$. By Theorem 4.5′ $\sigma$ is optimal in $\bar{C}$ for the stochastic sequence $\{x_n, \mathscr{H}_n\}_1^\infty$.

We show below that by paying particular attention to the "continuous time" aspects of the given problem we can eliminate some of the difficulties usually encountered in computing $\sigma$ and even carry out the computation explicitly in some particular cases.

For this purpose let

$$\mathscr{F}_u = \mathscr{B}(N(u'), u' \leq u; y_1, \ldots, y_{N(u)}). \qquad u \geq 0.$$

Then for each $n = 1, 2, \ldots$

$$\mathscr{H}_n = \mathscr{F}_{\tau_n}$$

where by definition

$$\mathscr{F}_{\tau_n} = \{A: A\{\tau_n \leq u\} \in \mathscr{F}_u \quad \text{for all} \quad u \geq 0\}.$$

Note that $\lim_{u\to\infty} EN(u) = \int_0^\infty p(u')\,du'$ may be finite. In this case there exists a first index $n$ (depending on the particular sample path $N(\cdot)$) such that $\tau_n = \infty$. By adopting the transformation of the time scale

$$\tilde{u} = \int_0^u p(u')\,du',$$

which maps the positive $u$-axis $1-1$ onto the finite or infinite interval $0 \leq \tilde{u} \leq \tilde{U} = \int_0^\infty p(u')\,du'$, we may assume $p(u) \equiv 1$. *Henceforth we shall omit the* $\tilde{}$*'s and with no loss in generality assume that* $p(u) \equiv 1$ *and that there exists a* $U \leq \infty$ *such that* $r$ *is positive on* $[0, U)$ *and zero on* $[U, \infty)$.

Putting $z_n = (y_n, \tau_n)$, we see that we are in the stationary Markov case. Moreover, $V_n(z) = \sup_t E(y_t r(u + \tau_t))$ is a function of $u$ only, say $V(u)$. Thus by Theorem 5.2

$$\sigma = \text{first } n \quad \text{such that} \quad y_n r(\tau_n) \geq V(\tau_n).$$

For $0 \leq u < U$ let $x_n(u) = y_n r(u + \tau_n)$ $(n = 1, 2, \ldots)$, and consider the family of stochastic sequences $\{x_n(u), \mathscr{F}_{\tau_n}\}_1^\infty$. Since for any fixed $u_0$ $P\{u_0 + \tau_n \,\varepsilon$ set of discontinuities of $r(\;)\} = 0$, it follows that $x_n(u) \downarrow x_n(u_0)$ as $u \uparrow u_0$ for any $0 < u_0 < U$ and $x_n(u) \uparrow x_n(u_0)$ as $u \downarrow u_0$ for any $0 \leq u_0 < U$. The remaining conditions of Theorem 4.9 are easily checked and we see that

$$V(u) = E\gamma_1(u)$$

is continuous on $[0, U)$. Putting

$$(5.12) \qquad y(u) = V(u)/r(u), \quad 0 \leq u < U,$$
$$= 0, \quad u \geq U,$$

we see that $y(\cdot)$ is piecewise continuous and that

$$\sigma(u) = \text{first } n \quad \text{such that} \quad y_n > y(u + \tau_n), \qquad (\sigma = \sigma(0)).$$

Denote

$$G(y) = 1 - F(y), \quad H(y) = \int_y^\infty y'\,dF(y'),$$

and for $0 \leq u \leq U$, $0 \leq v \leq U - u$ let

$$f_u(v) = P\{\tau_{\sigma(u)} > v\}.$$

Since $y(\cdot)$ is piecewise continuous we may apply the usual differential argument to obtain

$$P\{\tau_{\sigma(u)} > v + h \mid \tau_{\sigma(u)} > v\} = \frac{f_u(v + h)}{f_u(v)} = 1 - G(y(u + v))h + o(h),$$

which together with the analogous result for $h < 0$ leads to

$$\frac{f_u'(v)}{f_u(v)} = -G(y(u + v)).$$

Since $f_u(0) = 1$, we have

$$f_u(v) = \exp\left[-\int_u^{u+v} G(y(v'))\, dv'\right].$$

From

$$\begin{aligned} V(u) &= E[y_{\sigma(u)} r(u + \tau_{\sigma(u)})] \\ &= \int_0^{U-u} E(y_{\sigma(u)} r(u + \tau_{\sigma(u)}) \mid \tau_{\sigma(u)} = v)(-f_u'(v))\, dv \end{aligned}$$

we obtain after a change of variable

$$\text{(5.13)} \qquad y(u)r(u) = V(u) = \int_u^U r(v)H(y(v)) \times \exp\left[-\int_u^v G(y(v'))\, dv'\right] dv.$$

Differentiating with respect to $u$ we have

$$\text{(5.14)} \qquad \frac{d}{du}[r(u)y(u)] = -r(u)\varphi(y(u)),$$

where we have put

$$\text{(5.15)} \qquad \varphi(y) = H(y) - yG(y).$$

We now claim that:

(i) A necessary and sufficient condition that a piecewise continuous function $\tilde{y}(\cdot)$ satisfy (5.13) is that $\tilde{y}(\cdot)$ satisfy (5.14) and

$$\text{(5.16)} \qquad r(u)\tilde{y}(u) \to 0 \quad \text{as} \quad u \to U;$$

(ii) If $\tilde{y}(\cdot)$ satisfies (5.13) on $[0, U)$ and $\tilde{y}(\cdot) = 0$ on $[U, \infty)$, then

$$\sigma = \text{first } n \quad \text{such that} \quad y_n \geq \tilde{y}(\tau_n),$$

i.e., $\tilde{y}(\cdot)$ is the function $y(\cdot)$ of (5.12).

The proof of (i) is completely analytic and will be disposed of first. The necessity of (5.14) has been shown above. The necessity of (5.16) follows from (5.13) when we observe that exp $[\cdot] \leq 1$, $H(y) \leq \mu$ and apply (5.10).

Now suppose that $\tilde{y}(\cdot)$ satisfies (5.14) and (5.16), and define $y_1(\cdot)$ by

$$r(u)y_1(u) = \int_u^U r(u)H(\tilde{y}(u)) \exp\left[-\int_u^v G(\tilde{y}(v'))\, dv'\right] dv.$$

We wish to show that $y_1(\cdot) = \tilde{y}(\cdot)$. We first note that $r(u)y_1(u) \to 0$ as $u \to U$ for exactly the same reasons that (5.16) holds, and by differentiation we find that

$$(5.17) \qquad \frac{d}{du}[r(u)y_1(u) = -r(u)[H(\tilde{y}(u)) - y_1(u)G(\tilde{y}(u))].$$

Subtracting (5.14) from (5.17) we obtain

$$(5.18) \qquad \frac{d}{du}[r(u)(y_1(u) - \tilde{y}(u))] = G(\tilde{y}(u)) \cdot r(u)[y_1(u) - \tilde{y}(u)].$$

Assume that $y_1 \neq \tilde{y}$ at some point $u_0$ and let $u_1 \leq U$ be the first point after $u_0$ such that $r(y_1 - \tilde{y}) = 0$; if no such $u_1$ exists set $u_1 = U$. Integrating (5.18) we have for any $u_0 < u < u_1$

$$(5.19) \qquad \log[r(u)|y_1(u) - \tilde{y}(u)|] = \log[r(u_0)|y_1(u_0) - \tilde{y}(u_0)|] + \int_{u_0}^u G(\tilde{y}(u'))\, du'.$$

Since $r\tilde{y} \to 0$, $r\tilde{y}_1 \to 0$ as $u \to U$, we see on letting $u \to u_1$ that the left-hand side of (5.19) goes to $-\infty$ whereas the right-hand side is bounded away from $-\infty$. This contradiction completes the proof of (i).

To prove (ii) suppose there exists a function $\tilde{y}(u)$ vanishing on $[U, \infty)$ such that, putting $\tilde{V} = r\tilde{y}$ on $[0, U)$, we have for $0 \leq u \leq U$

$$(5.20) \qquad \tilde{V}(u) = \int_u^U r(v)H(\tilde{y}(v)) \exp\left[-\int_u^v G(\tilde{y}(v'))\, dv'\right] dv.$$

Defining

$$\tilde{\sigma}(u) = \text{first } n \quad \text{such that} \quad y_n r(u + \tau_n) \geq \tilde{V}(u + \tau_n)$$
$$(\tilde{\sigma} = \tilde{\sigma}(0))$$

and arguing as before, we see that the right-hand side of (5.20) is $E[y_{\tilde{\sigma}(u)}r(u + \tau_{\tilde{\sigma}(u)})]$. From well-known properties of the exponential distribution it follows that the conditional joint distribution given $\mathscr{F}_u$ of

$$(y_{N(u)+1}, \tau_{N(u)+1}), (y_{N(u)+2}, \tau_{N(u)+2}), \ldots$$

is the same as the unconditional joint distribution of

$$(y_1, u + \tau_1), (y_2, u + \tau_2), \ldots .$$

Observing that on $\{\tau_{\tilde{\sigma}} > u\}$

$$\tilde{\sigma} = \text{first } N(u) + n \quad \text{such that} \quad y_{N(u)+n} \geq \tilde{y}(\tau_{N(u)+n}),$$

we see that on $\{\tau_{\tilde{\sigma}} > u\}$ the conditional distribution given $\mathscr{F}_u$ of $y_{\tilde{\sigma}}r(\tau_{\tilde{\sigma}})$ is the same as the unconditional distribution of

$$y_{\tilde{\sigma}(u)}r(u + \tau_{\tilde{\sigma}(u)}).$$

Hence $\tilde{V}(u) = E(y_{\tilde{\sigma}}r(\tau_{\tilde{\sigma}}) \mid \mathscr{F}_u)$ on $\{\tau_{\tilde{\sigma}} > u\}$, or somewhat more generally

$$\text{(5.21)} \qquad \tilde{V}(u + v) = E(y_{\tilde{\sigma}(u)}r(\tau_{\tilde{\sigma}(u)}) \mid \mathscr{F}_v) \quad \text{on} \quad \{\tau_{\tilde{\sigma}(u)} > v\}$$

for all $0 \leq u < U$, $0 \leq v < U - u$. Now define $\tilde{\Gamma}(y, u) = \max(yr(u), \tilde{V}(u))$ for all $0 \leq y\infty$, $0 \leq u < U$. Conditional on $\mathscr{F}_{\tau_1}$ our expected reward if we use $\tilde{\sigma}(u)$ for $\{x_n(u), \mathscr{F}_{\tau_n}\}_1^\infty$ is

$$\begin{aligned} &y_1 r(u + \tau_1) \quad \text{if} \quad y_1 r(u + \tau_1) \geq \tilde{V}(u + \tau_1), \\ &E(y_{\tilde{\sigma}(u)}r(\tau_{\tilde{\sigma}(u)}) \mid \mathscr{F}_{\tau_1}) \quad \text{if} \quad y_1 r(u + \tau_1) < \tilde{V}(u + \tau_1), \end{aligned}$$

which by (5.21) is

$$\max(y_1 r(u + \tau_1), \tilde{V}(u + \tau_1)) = \tilde{\Gamma}(y_1, u + \tau_1).$$

Hence $\tilde{V}(u) = E\tilde{\Gamma}(y_1, u + \tau_1)$. Now define $\tilde{\gamma}_n = \tilde{\Gamma}(y_n, \tau_n)$ $(n = 1, 2, \ldots)$. Then

$$E(\tilde{\gamma}_{n+1} \mid \mathscr{F}_{\tau_n}) = E(\tilde{\Gamma}(y_{n+1}, \tau_{n+1}) \mid \mathscr{F}_{\tau_n}) = \tilde{V}(\tau_n)$$

and it follows from the definition of $\tilde{\Gamma}$ that

$$\text{(5.22)} \qquad \tilde{\gamma}_n = \max(x_n, E(\tilde{\gamma}_{n+1} \mid \mathscr{F}_{\tau_n})) \quad (n = 1, 2, \ldots).$$

From $\tilde{\gamma}_n \leq \sum_1^\infty x_k$ and (5.11) we have the existence of a random variable $X \geq 0$ such that $EX < \infty$ and

$$\text{(5.23)} \qquad \tilde{\gamma}_n \leq E(X \mid \mathscr{F}_{\tau_n}) \qquad (n = 1, 2, \ldots).$$

Finally since $x_n \to 0$ and $\tilde{V}(u) \to 0$ for the same reasons that imply (5.16) we have

$$\tilde{\gamma}_\infty = 0. \tag{5.24}$$

From (5.22), (5.23), (5.24), and Lemma 4.9 it follows that $\tilde{\gamma}_n \leq \gamma'_n$, $(n = 1, 2, \ldots)$. Moreover, by Theorem 4.6 $\{\gamma'_n, \mathscr{F}_{\tau_n}\}_1^\infty$ is the *minimal* stochastic sequence satisfying (5.22); and by Theorem 4.4 $(\gamma'_n) = (\gamma_n)$. It follows that $(\tilde{\gamma}_n) = (\gamma_n)$. This proves (ii).

**Examples**

(a) Suppose that $p(u) = 1$, $0 \leq u < U = \infty$, $r(u) = e^{-\alpha u}$. Then (5.14) becomes

$$\frac{dy}{du} - \alpha y = -\varphi(y);$$

and we find that $y(u) = y_0$, where $y_0$ is the unique root of

$$\alpha y = \varphi(y).$$

(See problem 5.2.)

(b) Now let us illustrate the case where $U < \infty$, i.e., where on the original time scale the expected number of offers is finite. Suppose that $r = 1$. Since for a relevant solution of (5.14) $y(U) = 0$, we obtain by integration

$$\int_0^y \frac{dy'}{\varphi(y')} = U - u. \tag{5.25}$$

The integral on the LHS is easily seen to diverge as $y$ approaches the (finite or infinite) upper bound of the distribution of $y_k$. Thus there exists a number $\eta$ such that

$$\int_0^\eta \frac{dy}{\varphi(y)} = U,$$

and to any $u$ in $[0, U)$ there is a unique solution of (5.25) with value in $(0, \eta)$. This solution is the critical curve. For example, if on the *original time scale* $p(u') = \lambda e^{-\lambda u'}$, and if $dF(y) = \theta e^{-\theta y}\, dy$, it is easily seen that

$$y(u') = \theta^{-1} \log (1 + e^{-\lambda u'}).$$

## 5. The Independent Case

If for each $n = 1, 2, \ldots$, $\mathscr{F}_n$ and $\mathscr{B}(x_{n+1})$ are independent, we say that we are in the independent case. In particular then $x_1, x_2, \ldots$ are independent random variables. The Secretary Problem, the Parking Problem, and the problem of Section 3.6(b) are independent case problems.

With $z_n = x_n$ $(n = 1, 2, \ldots)$ we have by Theorem 5.1 (see also the introductory comments of this chapter)

**Theorem 5.4** In the independent case for each $n \geq 1$
(a) $\gamma'_n = \max(x_n, v'_{n+1})$, $v'_n = E[\max(x_n, v'_{n+1})]$ (see (4.8) for definitions),
(b) $\gamma_n = \max(x_n, v_{n+1})$, $v_{n+1} = E[\max(x_n, v_{n+1})]$ (see (4.2′) for definitions),
(c) $v_n = \sup_{t \in D_n} Ex_t$,
(d) $\gamma_1, \gamma_2, \ldots$ are independent random variables.

For the remainder of this section we shall assume that *we are in the independent case and that* $V < \infty$. From Theorem 5.4(b) we have

$$\begin{aligned} \sigma &= \text{first } n \quad \text{such that} \quad x_n \geq v_{n+1} \\ &= \infty \quad \text{if no such } n \text{ exists.} \end{aligned}$$

**Lemma 5.3** $\lim_{n \to \infty} \gamma_n$ exists and $x_\infty = \gamma_\infty = v_\infty$.

***Proof*** By the Kolmogorov $0 - 1$ law for tail events (Section 2.2(b)) $x_\infty$ is a.s. constant. Suppose $x_\infty = -\infty$. By Lemma 4.13 $P\{x_n \geq v_{n+1} - \varepsilon, \text{i.o.}\} = 1$ for all $\varepsilon > 0$ and since $v_n \downarrow$, we have $v_n \to -\infty$. Thus $\gamma_n = \max(x_n, v_{n+1}) \to -\infty$. Now suppose that $x_\infty > -\infty$. By Theorem 4.7 $v_n \geq x_\infty$ $(n = 1, 2, \ldots)$ and thus $v_\infty \geq x_\infty$. But Lemma 4.13 now shows that $x_\infty = v_\infty$. Since by Theorem 5.4 $\gamma_n = \max(x_n, v_{n+1})$, for all $n = 1, 2, \ldots$, we see that $\lim_{n \to \infty} \gamma_n$ exists and $\gamma_\infty = x_\infty = v_\infty$.

**Theorem 5.5** If $x_\infty > -\infty$, then $\sigma$ is optimal in $\bar{C}$. If $x_\infty = -\infty$ and $Ex_\sigma$ exists, then $\sigma$ is optimal in $C$.

***Proof*** If $x_\infty > -\infty$, we have $x_\sigma \geq v_{\sigma+1} \geq v_\infty > -\infty$, since by Lemma 5.3 $v_\infty = x_\infty$. Thus we may assume that $Ex_\sigma$ exists, and since $V < \infty$, $Ex_\sigma < \infty$. The theorem now follows at once from Theorem 4.10(d) and the observation that by Theorem 5.4 $E(\gamma_{n+1} \mid \mathscr{F}_n) = v_{n+1} \leq v$ on $\{\sigma > n\}$ $(n \geq 1)$.

**Theorem 5.6** Let $\{\beta_n, \mathscr{F}_n\}_1^\infty$ be any stochastic sequence such that for each $n = 1, 2, \ldots$

(a) $$E\beta_n \text{ exists,}$$

(b) $$\beta_n = \max(x_n, E\beta_{n+1}),$$

(c) $P\{x_k \geq \beta_k - \varepsilon \quad \text{for some} \quad k \geq n\} = 1 \quad \text{for all} \quad \varepsilon > 0,$

and suppose that

(d) $$x_\infty > -\infty \quad \text{or} \quad E(\sup x_n^+) < \infty.$$

Then $\beta_n \leq \gamma_n$ $(n = 1, 2, \ldots)$. If $\limsup Ex_n > -\infty$, there exists precisely one sequence $\{\beta_n, \mathscr{F}_n\}_1^\infty$ satisfying (a), (b), and (c), to wit $(\beta_n) = (\gamma_n) = (\gamma_n')$.

***Proof*** Suppose that for some $n$ $P\{\beta_n > \gamma_n\} > 0$. It follows from condition (b) and Theorem 5.4(b) that $E\beta_{n+1} > v_{n+1}$ and thus by induction that $\beta_k \geq \gamma_k$, $k = n, n+1, \ldots$. Let $0 < \varepsilon < E\beta_n - v_n$. Let $t$ = first $k \geq n$ such that $x_k \geq \beta_k - \varepsilon$. By (c) $P(t < \infty) = 1$ and by (d) $E(x_t)$ and $E(\beta_t)$ exist. (If $x_\infty > -\infty$, we have $\beta_t \geq x_t \geq \beta_t - \varepsilon \geq \gamma_t - \varepsilon \geq v_\infty - \varepsilon = x_\infty - \varepsilon$.) By Lemma 3.3 and (b) we have

$$(5.26) \qquad E\beta_n = \int_{(t \leq k)} \beta_t + (E\beta_{k+1})P(t > k)$$

$$\leq \int_{(t \leq k)} x_t + \varepsilon + E^+(\beta_{k+1})P(t > k)$$

$$\to Ex_t + \varepsilon \quad \text{as} \quad k \to \infty.$$

Thus $Ex_t > v_n$, a contradiction. Now suppose that $\limsup Ex_n > -\infty$. Since $v_n \geq Ex_n$ and $x_\infty = v_\infty$, we have $x_\infty > -\infty$ and the first part of the theorem is applicable. By Theorems 4.6 and 5.4 $\{\gamma_n', \mathscr{F}_n\}_1^\infty$ is the minimal stochastic sequence satisfying (a) and (b). Thus it suffices to show that $(\gamma_n) = (\gamma_n')$. But since $v_n' \geq Ex_n$ and $v_n' \downarrow$ we have $\lim_{n\to\infty} v_n' = v_\infty' > -\infty$. Thus if $t \in C$

$$\int_{(t>n)} (\gamma_n')^- \leq (v_\infty')^- P(t > n) \to 0 \quad \text{as} \quad n \to \infty$$

and by Theorem 4.3,

$$\gamma_n' = \gamma_n \qquad (n = 1, 2, \ldots).$$

The following example shows that the second part of the above theorem is not true if we merely suppose that $x_\infty > -\infty$.

**Example**

Let $x_1, x_2, \ldots$ be independent, $\mathscr{F}_n = \mathscr{B}(x_1, \ldots, x_n)$ and suppose that

$$P\left\{x_n = -\prod_{k=1}^{n} 2^k\right\} = 2^{-n} = 1 - P\{x_n = 0\}$$

$$(n = 1, 2, \ldots).$$

Then $x_n \to 0$ (Borel-Cantelli), but $Ex_n \to -\infty$. Clearly $\sigma =$ first $n$ such that $x_n = 0$ is optimal in $C$. It is easily seen by backward induction, however, that $s^N = 1$, $Ex_{s^N} = -1$ for all $N = 1, 2, \ldots$.

## 6. The Independent Case–Applications

(a) Let $y_1, y_2, \ldots$ be i.i.d. with $E|y_1| < \infty$. Let $\mathscr{F}_n = \mathscr{B}(y_1, \ldots, y_n)$, $x_n = y_n - c_n$, where $(c_n)$ is any sequence of positive constants. If $c_{n+1} - c_n \uparrow$, then it follows from the results of Section 3.6(a) that $V < \infty$, and we shall show that $\sigma$ is optimal in $C$, $\sigma = \lim_{N\to\infty} s^N$. In fact from the assumption $c_{n+1} - c_n \uparrow$, we have $v_{n+1} + c_n \downarrow$. Hence as in Section 3.6(a) $\sigma$ has moments of all orders and $Ex_\sigma^+ < \infty$. That $\sigma$ is optimal in $C$ follows from Theorem 5.5, and an easy application of Theorem 4.4 and Lemma 4.6 shows that $\sigma = \lim_{N\to\infty} s^N$.

(b) Let $y, y_1, y_2, \ldots$ be i.i.d. with $E|y| < \infty$, and for each $n = 1, 2, \ldots$ let $\mathscr{F}_n = \mathscr{B}(y_1, \ldots, y_n)$ and $x_n = y_n/n$. From $E|y| < \infty$ it follows that for every $\varepsilon > 0$ $\sum_1^\infty P\{|y| > \varepsilon n\} < \infty$, and hence by the Borel-Cantelli lemma $x_n \to 0$. Thus without loss of generality we may assume that $P\{y \geq 0\} > 0$ (otherwise $V = 0$ and $\sigma = +\infty$ is optimal) and hence that $P\{y \geq 0\} = 1$. If $V < \infty$ (or equivalently by Theorem 4.14 if $E(y \log^+ y) < \infty$), then by Theorem 5.5 $\sigma$ is optimal in $\bar{C}$. If in addition $y \leq B < \infty$ or $E(y \mid y > a) = 0(a)$ as $a \to \infty$, then $\sigma \in C$. (See Section 5.6 for a discussion of this problem under different assumptions.)

To prove the last assertion, suppose that $E(y \mid y > a) = 0(a)$; the case where $y$ is bounded is similar. $\sigma =$ first $n$ such that $y_n \geq nv_{n+1}$. Since $nE\left(\dfrac{y_t}{n+t}\right) \leq (n+1)E\left(\dfrac{y_t}{n+1+t}\right)$ for any s.v. $t$, $nv_{n+1}$ is increasing. Since $x_\infty = 0$ and $\sigma$ is optimal in $\bar{C}$, we have

$$(5.27) \qquad V = \int_{(\sigma \leq n)} y_\sigma/\sigma + \int_{(n<\sigma<\infty)} y_\sigma/\sigma \qquad (n = 1, 2, \ldots).$$

It follows from (5.26) and (5.27) that for each $n$

$$v_{n+1}P(\sigma > n) = \int_{(n<\sigma<\infty)} y_\sigma/\sigma. \tag{5.28}$$

Now

$$\begin{aligned}\int_{(n<\sigma<\infty)} y_\sigma/\sigma &= \sum_{k=n+1}^{\infty} P(\sigma \geq k) \int_{(y_k \geq kv_{k+1})} y_k/k \\ &= \sum_{k=n+1}^{\infty} v_{k+1}P(\sigma = k)\,\frac{E(y \mid y \geq kv_{k+1})}{kv_{k+1}} \\ &\leq v_{n+1} \sup_k \frac{E(y \mid y \geq kv_{k+1})}{kv_{k+1}}\, P(n < \sigma < \infty).\end{aligned}$$

Since $kv_{k+1} \uparrow$ (and hence is bounded away from 0) we have by the assumption $E(y \mid y > a) = 0(a)$ that

$$\int_{(n<\sigma<\infty)} y_\sigma/\sigma \leq \text{const}\; v_{n+1} \quad P(n < \sigma < \infty). \tag{5.29}$$

It follows from (5.28) and (5.29) that $P\{\sigma > n\} \to 0$ as $n \to \infty$ and hence $\sigma \in C$.

## 7. Uniform Games

Let $y_1, y_2, \ldots$ be independent and uniform on $(0, 1)$, $\mathscr{F}_n = \mathscr{B}(y_1, \ldots, y_n)$, and for $\alpha > 0$ let the *loss* incurred if we stop at time $n$ be $x_n = n^\alpha y_n$ $(n = 1, 2, \ldots)$. We are interested in finding a s.v. $t$ which *minimizes* $Ex_t$.

### Theorem 5.7

(i) For any $0 < \alpha \leq 1$, $V = 0$ and $\sigma = +\infty$ is optimal in $\bar{C}$.

(ii) There exists a smallest number $\alpha^* > 1$ such that for any $\alpha \geq \alpha^*$, $V = \frac{1}{2}$ and $\sigma = 1$ is optimal in $C$.

(iii) For any $1 < \alpha < \alpha^*$

(a) $0 < V < \frac{1}{2}$,

(b) $\sigma$ is optimal in $C$,

(c) $E\sigma = \infty$,

(d) $\sigma = \lim_{N\to\infty} s^N$,

(e) $v_n \sim 2(\alpha - 1)n^{1-\alpha} \quad (n \to \infty)$.

(iv) $V$ is a continuous function of $\alpha$.

***Proof*** Let $A > 0$ be arbitrary

$$\sum^{\infty} P\{n^{\alpha}y_n \leq A\} = \sum^{\infty} n^{-\alpha}A. \tag{5.30}$$

Hence for any $\alpha \leq 1$ $P\{x_n \leq A, \text{i.o.}\} = 1$ by the Borel-Cantelli lemma, and since $A$ is arbitrary $x_{\infty} = 0$. This proves (i).

Let $u_n = \inf_{k \geq n} x_k$. For any $0 \leq A < n^{\alpha}$

$$P\{u_n \geq A\} = \prod_{k=n}^{\infty} \left(1 - \frac{A}{k^{\alpha}}\right). \tag{5.31}$$

Hence $P\{u_1 \geq \frac{1}{2}\} = \prod_{1}^{\infty} \left(1 - \frac{1}{2n^{\alpha}}\right) > 0$ $(\alpha > 1)$, so

$$V \geq Eu_1 > 0 \qquad (\alpha > 1). \tag{5.32}$$

By (5.30) and the Borel-Cantelli lemma we see that $x_{\infty} = \infty$ for all $\alpha > 1$ and hence by Theorem 4.5 $\sigma$ is optimal in $C$.

To complete the proofs of (ii)–(iv) we obtain bounds on $(v_n)$ from the functional equations of Theorem 5.4(b), which in this case become

$$v_n = \int_0^1 \min(n^{\alpha}y, v_{n+1})\, dy \qquad (n \geq 1). \tag{5.33}$$

From (5.33)

$$v_n \leq \tfrac{1}{2}n^{\alpha} \qquad (n \geq 1), \tag{5.34}$$

and thus

$$v_{n+1}n^{-\alpha} \leq \frac{1}{2}\left(\frac{n+1}{n}\right)^{\alpha} \leq \frac{1}{2}\left(\frac{3}{2}\right)^{\alpha} < 1 \qquad (n \geq 2,\ 1 < \alpha < \tfrac{3}{2}). \tag{5.35}$$

Hence (5.33) may be written as

$$v_n = v_{n+1}(1 - v_{n+1}/2n^{\alpha}) \qquad (n \geq 2,\ 1 < \alpha \leq \tfrac{3}{2}). \tag{5.36}$$

Since $v_n > 0$ for all $n = 1, 2, \ldots, \alpha > 1$ we can rewrite (5.36) as

$$\frac{1}{v_n} = \frac{1}{v_{n+1}} + \frac{1}{2n^{\alpha} - v_{n+1}} \qquad (n \geq 2,\ 1 < \alpha \leq \tfrac{3}{2}), \tag{5.37}$$

which on summing for $k = n, n+1, \ldots, m$ becomes

$$\frac{1}{v_n} = \sum_{k=n}^{m} \frac{1}{2k^{\alpha} - v_{k+1}} + \frac{1}{v_{n+1}} \qquad (n \geq 2,\ 1 < \alpha \leq \tfrac{3}{2}). \tag{5.38}$$

Since by Lemma 5.3 $v_\infty = x_\infty = \infty$ for all $\alpha > 1$ we may let $m \to \infty$ in (5.38) to obtain

$$\text{(5.39)} \qquad \frac{1}{v_n} = \sum_{k=n}^{\infty} \frac{1}{2k^\alpha - v_{k+1}} \qquad (n \geq 2, 1 < \alpha \leq \tfrac{3}{2}).$$

From (5.39) and (5.35) we obtain

$$\text{(5.40)} \qquad \frac{1}{2(\alpha - 1)(n+1)^{\alpha-1}} \leq \frac{1}{2}\int_{n+1}^{\infty} y^{-\alpha}\,dy \leq \frac{1}{2}\sum_{n+1}^{\infty} k^{-\alpha}$$
$$\leq \frac{1}{v_{n+1}} \leq \sum_{n+1}^{\infty} k^{-\alpha} \leq \frac{1}{(\alpha - 1)n^{\alpha-1}},$$

and hence that

$$\text{(5.41)} \qquad \frac{\alpha - 1}{n} \leq \frac{v_{n+1}}{n^\alpha} \leq \frac{2(\alpha-1)}{n+1}\left(\frac{n+1}{n}\right)^{\alpha} \qquad (n \geq 1, 1 < \alpha \leq \tfrac{3}{2}).$$

We now show that $v_2 > 1$ for $\alpha = \frac{3}{2}$. It follows from (5.33) that $V = \frac{1}{2}$ and $\sigma = 1$ for all $\alpha \geq \frac{3}{2}$. From (5.36)

$$\text{(5.42)} \qquad v_{n+1} = n^\alpha - \sqrt{n^{2\alpha} - 2n^\alpha v_n} \qquad (n \geq 2, 1 < \alpha \leq \tfrac{3}{2}),$$

the $+$ sign being excluded by (5.34). Suppose that $\alpha = \frac{3}{2}$ and $v_2 \leq 1$. Then by (5.42)

$$v_3 \leq 2^{3/2} - \sqrt{8 - 2 \cdot 2^{3/2}} = 1.3$$
$$\text{(5.43)} \qquad v_4 \leq 1.52$$
$$v_5 \leq 1.7.$$

On the other hand, by (5.41)

$$\frac{v_{n+1}}{n^\alpha} \leq \frac{1}{n+1}\left(\frac{n+1}{n}\right)^{3/2} \leq \frac{4}{50} \qquad (n \geq 5).$$

Hence from (5.39) with $\alpha = \frac{3}{2}$

$$\frac{1}{v_5} = \sum_{5}^{\infty} \frac{1}{2k^\alpha - v_{k+1}} \leq \sum_{5}^{\infty} \frac{1}{2k^\alpha(1 - \frac{11}{100})} \leq \frac{50}{89}\int_{9/2}^{\infty} t^{-\alpha}\,dt < \frac{1}{1.7},$$

contradicting (5.43). Hence $V = \frac{1}{2}$ and $\sigma = 1$ if $\alpha \geq \frac{3}{2}$.

By (5.41) $V \to 0$ as $\alpha \to 1$. Let $\alpha$ be any number $>1$ for which $V < \frac{1}{2}$. Then by (5.33) $P\{\sigma > 1\} > 0$, and since

$$\sigma = \inf\{n: y_n \leq v_{n+1}/n^\alpha\},$$

it follows from (5.35) that $P\{\sigma > N\} > 0$ for every $N$. Moreover, if $1 < \alpha = \frac{3 - \varepsilon}{2}$ for some $\varepsilon > 0$, by (5.41) we have

$$v_{n+1}/n^{\alpha} \leq (1 - \varepsilon)\left(\frac{n + 1}{n^2}\right) \leq \frac{1}{n} \quad \text{for} \quad n \geq (1 - \varepsilon)/\varepsilon.$$

Hence if $V < \frac{1}{2}$, so $\sigma$ is unbounded, we have for any $n > N \geq (1 - \varepsilon)/\varepsilon$ and some $K > 0$,

$$P\{\sigma > n\} \geq K(1 - 1/N)\left(1 - \frac{1}{N + 1}\right)\cdots(1 - 1/n)$$
$$= K\frac{N - 1}{n},$$

so $E\sigma = \sum_0^{\infty} P\{\sigma > n\} = \infty$. Thus for any $\alpha > 1$ either $V = \frac{1}{2}$ and $\sigma = 1$, or $0 < V < \frac{1}{2}$, $P\{\sigma < \infty\} = 1$, but $E\sigma = \infty$.

To prove (iii)(e) we first observe that for any $1 < \alpha < \frac{3}{2}$ for which $V < \frac{1}{2}$, $\frac{v_{n+1}}{n^{\alpha}} \to 0$ $(n \to \infty)$ by (5.41). Hence for any $\varepsilon > 0$, $v_{k+1} \leq \varepsilon k^{\alpha}$ for all $k$ sufficiently large, and thus from (5.39) we have for all $n$ sufficiently large

$$\frac{1}{v_n} = \sum_{k=n}^{\infty} \frac{1}{2k^{\alpha} - v_{k+1}} \leq \frac{1}{2 - \varepsilon}\sum_n^{\infty} k^{-\alpha} \leq \frac{1}{2 - \varepsilon}\int_{n-1}^{\infty} x^{-\alpha}\, dx$$
$$= \frac{1}{(2 - \varepsilon)(\alpha - 1)(n - 1)^{\alpha - 1}}.$$

It follows that $\liminf_{n\to\infty} \frac{v_n}{n^{\alpha-1}} \geq (2 - \varepsilon)(\alpha - 1)$, which since $\varepsilon > 0$ is arbitrary combines with (5.41) to give (iii)(e).

We now turn to the proof of (iii)(d) and (iv). ((iv) in turn completes the proof of (ii) by showing that $V = \frac{1}{2}$ for $\alpha = \alpha^*$.) The same argument that led to (5.41) shows that

$$\frac{\alpha - 1}{n} \leq \frac{v'_{n+1}}{n^{\alpha}} \leq \frac{2(\alpha - 1)}{n + 1}\left(\frac{n + 1}{n}\right)^{\alpha} \qquad (n \geq 1,\ 1 < \alpha \leq \tfrac{3}{2}). \tag{5.44}$$

By Theorem 4.10 and (5.41) we have

$$(\alpha - 1)n^{\alpha-1}P\{\sigma > n\} \leq v_{n+1}P\{\sigma > n\} = \int_{(\sigma > n)} \gamma_n \to 0$$
$$\text{as} \quad n \to \infty.$$

Hence by (5.44)

$$\int_{(\sigma>n)} \gamma_n' = v_{n+1}' P\{\sigma > n\} \leq \text{const } n^{\alpha-1} P\{\sigma > n\} \to 0$$

as $n \to \infty$, and by the remark following Theorem 4.3 and Lemma 4.6 we have $V' = V$, $\sigma = \lim_{N\to\infty} s^N$. The continuity of $V$ from the right at $\alpha > 1$ now follows from Theorem 4.9(b). (Continuity from the right at $\alpha = 1$ is immediate from (5.41).) That $V$ is left-continuous at any $\alpha > 1$ follows from Theorem 4.9(a). This completes the proof.

## 8. The Problem of $y_n/n$

Let $y_1, y_2, \ldots$ be i.i.d. non-negative random variables with $Ey_1 = 1$. Let $\mathscr{F}_n = \mathscr{B}(y_1, \ldots, y_n)$ and $x_n = y_n/n$ $(n = 1, 2, \ldots)$. In Section 5.6(b) it was shown that if $E(y_1 \log^+ y_1) < \infty$ and $E(y_1 \mid y_1 > a) = 0(a)$ as $a \to \infty$, then $\sigma$ is optimal in $C$. In this section we prove the same result under the hypothesis that for some $\alpha > 1$

$$Ey_1^\alpha < \infty. \tag{5.45}$$

First note that (5.45) implies that

$$\begin{aligned} v_n &\leq E(\sup_{k\geq n} x_k) \leq (E(\sup_{k\geq n} x_k^\alpha))^{1/\alpha} \\ &\leq \text{const}\left(\sum_n^\infty k^{-\alpha}\right)^{1/\alpha} = 0(n^{-1+1/\alpha}), \end{aligned} \tag{5.46}$$

and hence by Theorem 4.5′ $\sigma$ is optimal in $\bar{C}$. To show that $\sigma \in C$ it suffices to show that $P\{\sigma = \infty\} > 0$ implies

$$\frac{v_{2n}}{v_n} \to 1 \quad \text{as} \quad n \to \infty. \tag{5.47}$$

For if (5.47) holds, then for any $\varepsilon > 0$ we have by (5.46) for all $m$ sufficiently large and all $n = 1, 2, \ldots$

$$\text{const } (2^n m)^{-1+1/\alpha} \geq v_{2^n m} \geq (1 - \varepsilon)^n v_m$$

and hence

$$\log v_m + n \log (1 - \varepsilon) \leq (1/\alpha - 1)(n \log 2 + \log m) + 0(1).$$

Letting $n \to \infty$, we obtain

$$\log (1 - \varepsilon) \leq (1/\alpha - 1) \log 2,$$

a contradiction for $\varepsilon$ sufficiently small.

Suppose then that $P\{\sigma = \infty\} > 0$. Since $x_\infty = 0$, Theorem 4.5′ applied to the stochastic sequence $\{x_k, \mathscr{F}_k\}_n^\infty$ shows that for each $n = 1, 2, \ldots$

$$v_n = \sum_{i=0}^{\infty} (n+i)^{-1} \int_{\{\sigma_n = n+i\}} y_{n+i}. \tag{5.48}$$

Put

$$p_n = P\{y_n < nv_{n+1}\}. \tag{5.49}$$

By Theorem 5.4 and the definition of $\sigma_n$ we have $P\{\sigma_n = n+i\} = p_n p_{n+1} \cdots p_{n+i-1}(1 - p_{n+i})$ and (5.48) becomes

$$v_n = \sum_{i=0}^{\infty} (p_n \cdots p_{n+i-1})(n+i)^{-1} \int_{\{y_{n+i} \geq (n+i)v_{n+i+1}\}} y_{n+i}. \tag{5.50}$$

Now define a new stopping rule $t$ by

$$t = \inf\{m : m \geq 2n, y_m \geq [m/2]v_{[m/2]+1}\}, \tag{5.51}$$

where $[x]$ denotes the integral part of $x$. Then $t \in \bar{C}_{2n}$ and hence

$$v_{2n} \geq Ex_t = \sum_{m=2n}^{\infty} m^{-1} \int_{\{t=m\}} y_m. \tag{5.52}$$

But by (5.49), (5.51), and the independence of the $y$'s we have for each $i = 0, 1, \ldots$

$$\int_{\{t=2n+2i\}} y_{2n+2i} = p_n^2 \cdots p_{n+i-1}^2 \int_{\{y_{2n+2i} \geq (n+i)v_{n+i+1}\}} y_{2n+2i}$$

and

$$\int_{\{t=2n+2i+1\}} y_{2n+2i+1} = p_n^2 \cdots p_{n+i-1}^2 p_{n+i} \int_{\{y_{2n+2i+1} \geq (n+i)v_{n+i+1}\}} y_{2n+2i+1}.$$

Since the $y$'s have a common distribution, the integrals on the right-hand side of the last two equations each equal

$$\int_{\{y_{n+i} \geq (n+i)v_{n+i+1}\}} y_{n+i}.$$

Substituting in (5.52) we obtain

$$Ex_t = \sum_{i=0}^{\infty} (p_n^2 \cdots p_{n+i-1}^2 (2n+2i)^{-1} + p_n^2 \cdots p_{n+i-1}^2 p_{n+i}(2n+2i+1)^{-1}) \int_{\{y_{n+i} \geq (n+i)v_{n+i+1}\}} y_{n+i}. \tag{5.53}$$

From this, (5.50), and (5.52) we obtain

$$\frac{v_{2n}}{v_n} \geq \inf_{i \geq 0} (p_n \cdots p_{n+i-1}(2n + 2i)^{-1} + p_n \cdots p_{n+i}(2n + 2i + 1)^{-1})/(n + i)^{-1}$$

$$\geq \left(\prod_{i=0}^{\infty} p_{n+i}\right) \inf_{i \geq 0} \frac{(n + i)^2}{(n + i + \frac{1}{2})^2}. \tag{5.54}$$

But $\prod_{n=1}^{\infty} p_n = P\{\sigma = \infty\}$, and thus if $P\{\sigma = \infty\} > 0$, the right-hand side of (5.54) tends to 1 as $n \to \infty$. But this yields (5.47), completing the proof.

## PROBLEMS

**1.** Complete problem 2 of Chapter 3, i.e., show that $\sigma$ is optimal and has the property that for some positive $a$, $\sigma =$ first $n$ such that $y_1 + \cdots + y_n \geq a$.

**2.** In the problem of Section 5.4, assume that $p(u) = 1, 0 \leq u < U = \infty$, $r(u) = e^{-\alpha u}$. Find and solve an equivalent monotone case problem.

**3.** Show that if $c_{n+1} - c_n$ is decreasing, the problems of Sections 3.6(a) and 5.6(a) are actually the same. What is the significance of this observation for purposes of computing $\sigma$ by backward induction and passing to the limit?

**4.** Prove directly that there exists exactly one sequence of constants tending to $\infty$ and satisfying (5.39). This gives an alternative proof of (ii)(d) and (iii) of Section 5.7.

**5.** Consider the problem of Sections 3.1(g) and 5.2(a). Describe the optimal rule $\sigma$ if the cost of sampling depends on the true density, i.e., if

$$-x_n = \min(a\pi_n, b\bar{\pi}_n) + c_0 n\pi_n + c_1 n\bar{\pi}_n \qquad (\bar{\pi}_n = 1 - \pi_n).$$

**6.** *Continuation.* Formulate this problem as an optimal stopping problem on $(\Omega, \mathcal{F}, P_1)$ along the lines of Section 5.2(c).

**7.** *Continuation.* What is the relation as $c_0 \to 0$, $b \to \infty$ of this problem to that of Section 5.2(c)?

**8.** Prove the lemma of Section 5.2(d).

**9.** Formulate and prove a version of Theorem 4.5′ appropriate to 2(d).

**10.** Apply the results of problems 8 and 9 to the optimal stopping problem of 5.2(d).

**11.** Let $y_1, y_2, \ldots$ be independent with $P(y_k = 1) = p = 1 - P(y_k = 0)$ $(k = 1, 2, \ldots)$. Let

$$x_n = \sum_1^n I_{\{y_k = y_{k+1} = \cdots = y_n = 1\}} - cn \quad \text{and} \quad \mathscr{F}_n = \mathscr{B}(y_1, \ldots, y_n)$$

for some $0 < c < 1$. (The sum appearing in the definition of $x_n$ denotes the length of the run of 1's counting backward from the $n$th observation.) Find an optimal stopping rule for $\{x_n, \mathscr{F}_n\}_1^\infty$. What is $V$? (This problem was suggested by N. Starr.)

**12.** (See Taylor [1].) Let $y_1, y_2,$ be i.i.d. with $-\infty < Ey_1 < 0$ and $E[(y_1^+)^2] < \infty$. Let $\mathscr{F}_n = \mathscr{B}(y_1, \ldots, y_n)$ and $x_n = \left(\sum_1^n y_k\right)^+$. (By Theorems 4.5′ and 4.13 $\sigma$ is optimal in $\bar{C}$ for $\{x_n, \mathscr{F}_n\}_1^\infty$.) Show that

$$\begin{aligned} \sigma &= \text{first } n \geq 1 \quad \text{such that} \quad x_n \geq q^{-1}Ex_\tau, \\ &= \infty \quad \text{if no such } n \text{ exists,} \end{aligned}$$

where

$$\begin{aligned} \tau &= \text{first } n \geq 1 \quad \text{such that} \quad x_n > 0 \\ &= \infty \quad \text{if no such } n \text{ exists} \end{aligned}$$

and $q = P(\tau < \infty)$. (*Hint:* This problem may be solved directly using Theorem 5.2 or may be reduced to the burglar problem of Chapter 3—see problems 3.2 and 5.1.)

# Bibliographical Notes

## Chapter 1

Chapter 1 contains a brief description of the measure-theoretic foundations of probability as proposed by Kolmogorov [1]. For a more detailed treatment, see (for example) Loève [1]. Lemma 1.2 is slightly more general than the usual form of Fatou's lemma. This generalization is used in Chapter 4 (e.g., Lemma 4.8, Theorem 4.10). Theorem 1.4 is due to Lévy [1], p. 129.

## Chapter 2

The fundamental results of martingale theory, in particular Theorems 2.1 and 2.2, are due to Doob [1], [2]. Parts (b) and (c) of Theorem 2.3 were proved by Chow, Robbins, and Teicher [1]. The idea to use martingales to obtain the ballot theorem of L. Takács (Section 4.4 (c)) is due to G. Simons.

Section 2.5 is taken from Siegmund [1], [4], although the basic ideas may be found throughout the recent literature of probability and statistics (see, e.g., Chow and Robbins [6] and Gundy and Siegmund [1]).

The circle of ideas surrounding Sections 2.6 and 2.7 originated with Wald (e.g., [1]). Use of the measure $Q$ is due to Bahadur [1].

## Chapter 3

Theorem 3.1 and Lemma 3.1 were proved by Robbins and Samuel [1].

The Bayes property of the sequential probability ratio test (Example

3.1 (h)) was discussed originally by Wald and Wolfowitz [1] and by Arrow, Blackwell, and Girshick [1] and more recently by Chow and Robbins [3] and Siegmund [2]. Arrow et al. recognized the optimal stopping aspect of the problem, which they solved by a finite case approximation. In particular, Theorem 3.2 originates with them.

The Secretary Problem (Example 3.1 (t)) has been studied by many authors. The paper by Gilbert and Mosteller [1] contains a bibliography of much of this work as well as discussions of several variations of the problem. Chow, Moriguti, Robbins, and Samuels [1] have studied the problem of minimizing the expected rank of the girl selected and have shown that

$$\lim_{N\to\infty} V^N = \prod_{j=1}^{\infty}\left(\frac{j+2}{j}\right)^{1/(j+1)} \cong 3.8695.$$

The basic results of Section 3.4 are due to Chow and Robbins [2], who isolated and studied the monotone case (in particular the problem of Section 3.6 (a)). Independently Derman and Sacks [1] studied a similar problem by similar methods. The idea of applying Lemma 3.1 to the problem of Section 3.6 (a) is due to D. Burdick. Other contributors to this problem and numerous variations of it include MacQueen and Miller [1], Sakaguchi [1], Bramblett [1], Yahav [1], Cohn [1], and DeGroot [1].

The problems of 3.6 (c) and (d) were studied by Wald [2] and by Mallows and Robbins [1] respectively.

## Chapter 4

The general theory of Chapter 4 is taken largely from Chow and Robbins [3], [5] and Siegmund [2]. Arrow et al. [1], Snell [1], and Haggstrom [1] have also studied the foundations of optimal stopping theory, and versions of several of our main theorems may be found in the work of these authors.

The existence of an optimal rule under the conditions of Theorem 4.5 was proved by Snell [1], who generalized an earlier result of Arrow et al. [1]. Chow and Robbins [3] gave a different proof of this result. Independently Haggstrom [1] and Chow and Robbins [1] identified the optimal rule as $\sigma$. Theorem 4.5′ is due to Siegmund [2]. Lemma 4.6 and its systematic application are new.

Lemma 4.7 was proved by Dvoretzky (see Bramblett [1]). The proof given is new.

The problem of Section 4.3 (e) has been studied in a series of papers by Shiryaev (cf. [1], [2]).

The concept of regularity of a supermartingale (Section 4.4) is due to Snell [1] and forms the basis for his study of optimal stopping problems.

Theorem 4.10 is new.

The optimal stopping problem for $s_n/n$ was first studied by Chow and Robbins [4], who proved the existence of an optimal stopping rule when the $y$'s assume the values $\pm 1$ with probabilities 1/2 each. Dvoretzky generalized this result to the case in which the $y$'s are independent and identically distributed with mean 0 and finite second moment. The discussion given also borrows from Siegmund, Simons, and Feder [1] and Ruiz-Moncayo [1]. Recently Shepp [1] has obtained more explicit information about the optimal rule $\sigma$ by relating the given problem to an analogous problem for the Wiener process, for which an exact solution may be obtained.

Theorem 4.12 (when $x_n \geq 0$ for all $n$) and its corollary are due to Samuel [1]. Theorem 4.13 may be distilled from the work of Chow and Robbins [2], Kiefer and Wolfowitz [1], and also from unpublished work of Darling, Erdös, and Kakutani. The proof given here borrows from the method of Erdös [1]. The proof in Theorem 4.14 that (a) implies (b) is due to Doob [1]. The converse part of the theorem was obtained independently by B. Davis [1] and by McCabe and Shepp [1].

## Chapter 5

The ideas behind the Markov case are implicit in the work of various authors. The discussion presented is taken from Siegmund [2], who also proved the closely related Theorem 5.3 (see also Dynkin and Yushkevich [1]).

The problem of Section 5.4 was studied by Elfving [1], who assumed that there is an optimal rule defined in terms of a piecewise continuous function $y(\cdot)$, derived the basic integral equation (5.13), and proved (i). Siegmund [2] removed Elfving's assumptions.

The independent case was first studied by Chow and Robbins [5], and later by Chow [1] and Siegmund [3].

The results of Section 5.7 are taken from Chow and Robbins [1], those of Section 8 from Chow and Dvoretzky [1].

# Bibliography

**Arrow, K. J., Blackwell, D., and Girshick, M. A.**

[1] Bayes and minimax solutions of sequential decision problems, *Econometrica*, **17** (1949), 213–244.

**Bahadur, R. R.**

[1] A note on the fundamental identity of sequential analysis, *Ann. Math. Statist.* **29** (1958), 534–543.

**Bellman, Richard**

[1] A problem in the sequential design of experiments, *Sankhyā* **16** (1956), 221–229.

**Bramblett, J. E.**

[1] Some approximations to optimal stopping procedures, Ph.D. dissertation (unpublished), New York: Columbia University, 1965.

**Breiman, L.**

[1] "Stopping-rule problems" in *Applied Combinatorial Mathematics*, E. F. Beckenbach (Editor), New York: John Wiley & Sons, Inc., 1964.

**Burkholder, D. L. and Gundy, R. F.**

[1] Extrapolation and interpolation of quasi-linear operators on martingales, *Acta Mathematica* **24** (1970), 249–304.

**Chernoff, H.**

[1] A note on risk and maximal regular generalized submartingales in stopping problems, *Ann. Math. Statist.* **38** (1967), 606–607.

[2] Optimal stochastic control, *Sankhyā*, Series A, **30** (1968), 221–252.

**Chow, Y. S.**

[1] On optimal stopping rules for independent random variables, mimeographed, Lafayette, Ind.: Purdue University, 1966.

**Chow, Y. S. and Dvoretzky, A.**

[1] Stopping rules for $x_n/n$ and related problems, Stanford, Calif.: Stanford Univ. Tech. Report, 1969.

**Chow, Y. S., Moriguti, S., Robbins, H., and Samuels, S. M.**

[1] Optimal selection based on relative rank, *Israel Journal of Mathematics* **2** (1964), 81–90.

**Chow, Y. S. and Robbins, H.**

[1] A class of optimal stopping problems, *Proc. Fifth Berkeley Symposium Math. Statist. Prob.*, Vol. 1, Berkeley and Los Angeles: University of California Press, 1967, 419–426.

[2] A martingale system theorem and applications, *Proc. Fourth Berkeley Symposium Math. Statist. Prob.*, Vol. 1, Berkeley and Los Angeles: University of California Press, 1961, 93–104.

[3] On optimal stopping rules, *Z. Wahrscheinlichkeitstheorie* **2** (1963), 33–49.

[4] On optimal stopping rules for $S_n/n$, *Illinois J. Math.* **9** (1965), 444–454.

[5] On values associated with a stochastic sequence, *Proc. Fifth Berkeley Symposium Math. Statist. Prob.*, Vol. 1, Berkeley and Los Angeles: University of California Press, 1967, 427–440.

[6] A renewal theorem for random variables which are dependent or non-identically distributed, *Ann. Math. Statist.* **36** (1963), 457–462.

**Chow, Y. S., Robbins, H., and Teicher, H.**

[1] Moments of randomly stopped sums, *Ann. Math. Statist.* **36** (1965), 789–799.

**Chung, K. L. and Fuchs, W. H. J.**

[1] On the distribution of values of sums of random variables, *Memoirs Amer. Math. Soc.* **6** (1950).

**Cohn, H.**

[1] On certain optimal stopping rules, *Rev. Roumaine Math. Pures Appl.* **12** (1967), 1173–1177.

**Davis, B.**

[1] Stopping rules for $S_n/n$ and the class $L \log L$, New Brunswick, N.J.: Rutgers Univ. Tech. Report, 1969.

**De Groot, M.**

[1] Some problems of optimal stopping, *Roy. Stat. Soc. B* **30** (1968), 108–122.

[2] *Optimal Statistical Decisions*, New York: McGraw-Hill, Inc., 1970.

**Derman, C. and Sacks, J.**

[1] Replacement of periodically inspected equipment (an optimal stopping rule), *Naval Research Logistics Quarterly* **7** (1960), 597–607.

**Doob, J. L.**

[1] Regularity properties of certain families of chance variables, *Trans. Amer. Math. Soc.* **47** (1940), 455–486.

[2] *Stochastic Processes*, New York: John Wiley & Sons, Inc., 1953.

**Dubins, L. E. and Teicher, H.**

[1] Optimal stopping when the future is discounted, *Ann. Math. Statist.* **38** (1967), 601–605.

**Dvoretzky, A.**

[1] Existence and properties of certain optimal stopping rules, *Proc. Fifth Berkeley Symposium Math. Statist. Prob.*, Berkeley and Los Angeles: University of California Press, 1967, 441–452.

**Dynkin, E. B.**

[1] The optimum choice of the instant for stopping a Markov process, *Doklady Akad. Nauk SSS* **150** (1963), 238–240; *Soviet Math.* **4** (1963), 627–629.

**Dynkin, E. B. and Yushkevich, A. A.**

[1] *Markov Processes, Theorems and Problems*, New York: Plenum Press, 1969.

**Elfving, G.**

[1] A persistency problem connected with a point process, *J. Appl. Prob.* **4** (1967), 77–89.

**Erdös, P.**

[1] On a theorem of Hsu and Robbins, *Ann. Math. Statist.* **20** (1949), 286–291.

**Ferguson, T.**

[1] *Mathematical Statistics, a Decision Theoretic Approach*, New York: Academic Press, Inc., 1967.

**Gilbert, J. and Mosteller, F.**

[1] Recognizing the maximum of a sequence, *Amer. Stat. Assoc.* **61** (1966), 35–73.

**Gundy, R. F. and Siegmund, D.**

[1] On a stopping rule and the central limit theorem, *Ann. Math. Statist.* **38** (1967), 1915–1917.

**Haggstrom, G.**

[1] Optimal stopping and experimental design, *Ann. Math. Statist.* **37** (1966), 7–29.

**Kiefer, J. and Wolfowitz, J.**

[1] On the theory of queues with many servers, *Trans. Amer. Math. Soc.* **78** (1955), 1–18.

**Kolmogorov, A. N.**

[1] Grundbegriffe der Wahrscheinlichkeitsrechnung, *Ergebnisse der Mathematik* **2** (1933), No. 3. (English translation: *Foundations of Probability Theory*, New York: Chelsea, 1955.)

**Lehmann, E. L.**

[1] *Testing Statistical Hypotheses*, New York: John Wiley & Sons, Inc., 1959, 105.

**Lévy, P.**

[1] *Theorie de l'Addition des Variables Aléatoires*, Paris: Gauthier-Villars, 1937.

**Loève, M.**

[1] *Probability Theory*, 3rd ed., Princeton, N.J.: D. Van Nostrand Co., Inc., 1963.

**McCabe, B. J. and Shepp, L. A.**

[1] On the supremum of $S_n/n$, *Ann. Math. Statist.* **41** (1970), 2166–2168.

**MacQueen, J. and Miller, R. G., Jr.**

[1] Optimal persistence policies, *Operat. Res.* **8** (1960), 362–380.

**Mallows, C. and Robbins, H.**

[1] Some problems of optimal sampling strategy, *J. Math. Anal. Appl.* **8** (1964), 90–103.

**Neveu, J.**

[1] *Mathematical Foundations of the Calculus of Probability*, San Francisco: Holden-Day, 1965, 44.

**Robbins, H. E. and Samuel, E.**

[1] An extension of a lemma of Wald, *J. Appl. Prob.* **3** (1966), 272–273.

**Ruiz-Moncayo, A.**

[1] Optimal stopping for functions of Markov chains, *Ann. Math. Statist.* **39** (1968), 1905–1912.

**Sakaguchi, M.**

[1] Dynamic programming of some sequential sampling designs, *J. Math. Anal. Appl.* **2** (1961), 446–466.

**Samuel, E.**

[1] On the existence of the expectation of randomly stopped sums, *J. Appl. Prob.* **4** (1967), 197–200.

**Shepp, L.**

[1] Explicit solutions to some problems of optimal stopping, *Ann. Math. Statist.* **40** (1969), 993–1010.

**Shiryaev, A. N.**

[1] On Markov sufficient statistics in non-additive Bayes problems of sequential analysis, *Theory Prob. Appl.* **9** (1964), 604–618.

[2] On optimal methods in quickest detection problems, *Theory Prob. Appl.* **8** (1963), 22–46.

[3] *Statistical Sequential Analysis*, Moscow, 1969. (In Russian.)

**Siegmund, D.**

[1] Some one-sided stopping rules, *Ann. Math. Statist.* **38** (1967), 1641–1646.

[2] Some problems in the theory of optimal stopping rules, *Ann. Math. Statist.* **38** (1967), 1627–1640.

[3] Some problems in the theory of optimal stopping rules, mimeographed, New York: Columbia University, 1966.

[4] The variance of one-sided stopping rules, *Ann. Math. Statist.* **40** (1969), 1074–1077.

**Siegmund, D., Simons, G., and Feder, P.**

[1] Existence of optimal stopping rules for rewards related to $S_n/n$, *Ann. Math. Statist.* **39** (1968), 1228–1235.

**Snell, J. L.**

[1] Application of martingale system theorems, *Trans. Amer. Math. Soc.* **73** (1952), 293–312.

**Taylor, H.**

[1] Bounds on the expected value of the positive part of a stopped random sum, unpublished manuscript, 1970.

**Teicher, H. and Wolfowitz, J.**

[1] Existence of optimal stopping rules for linear and quadratic rewards, *Z. Wahrscheinliehkeitstheorie* **5** (1966), 316–368.

**von Bahr, B. and Esseen, C. G.**

[1] Inequalities for the $r$th absolute moment of a sum of random variables, $1 \leq r \leq 2$, *Ann. Math. Statist.* **36** (1965), 299–303.

**Wald, A.**

[1] *Sequential Analysis*, New York: John Wiley & Sons, Inc., 1947.

[2] *Statistical Decision Functions*, New York: John Wiley & Sons, Inc., 1950.

**Wald, A. and Wolfowitz, J.**

[1] Optimum character of the sequential probability ratio test, *Ann. Math. Statist.* **19** (1948), 326–329.

**Walker, L. H.**

[1] Regarding stopping rules for Brownian motion and random walks, *Bull. Amer. Math. Soc.* **75** (1969), 46–50.

**Yahav, J. A.**

[1] On optimal stopping, *Ann. Math. Statist.* **37** (1966), 30–35.

# List of Symbols

(Page number refers to the definition of the notation.)

# Index

A CATALOG OF SELECTED

# DOVER BOOKS

IN SCIENCE AND MATHEMATICS